The dating of rocks and related isotope studies have become increasingly important branches of the Earth Sciences. A large gap now exists between those who obtain and interpret results, and the field geologist who wishes to apply these results to conventional geological problems. This book now fills the gap. For a fuller appreciation of results the principles of the physical and chemical methods are described in some detail. The results are interpreted in terms of various problems encountered by the geologist in routine field studies. Apart from dating the age of rocks and minerals, related isotope studies can, in many cases, be applied to fundamental problems of petrology and geophysics.

APPLIED
GEOCHRONOLOGY

APPLIED GEOCHRONOLOGY

E. I. HAMILTON
Department of Geology and Mineralogy
University Museum
Oxford
England

With a Chapter on Comparative Geochemistry

by

L. H. AHRENS
Department of Geochemistry
University of Capetown
South Africa

1965
ACADEMIC PRESS
London and New York

ACADEMIC PRESS INC. (LONDON) LTD
Berkeley Square House
Berkeley Square
London, W.1.

U.S. Edition published by

ACADEMIC PRESS INC.
111 Fifth Avenue
New York, New York 10003

Library of Congress Catalog Card Number: 65–18444

PRINTED IN GREAT BRITAIN AT
THE UNIVERSITY PRESS
ABERDEEN

PREFACE

The dating of rocks and the related field of isotope abundance studies have become increasingly important branches of the Earth Sciences. The technical aspects of this work and the interpretation of the results are specialized. A rather wide gap now exists between those who obtain the results and the geologists who wish to apply them to conventional geological problems. The whole fields of radiometric dating and of isotope abundance studies, which stem from the pioneering work of A. O. Nier and A. Holmes, are in a state of constant growth punctuated by periods of rapid expansion. In the past, books describing various aspects of dating have been available.*

This volume is written specifically for geologists, but it is hoped that others interested in the general field of the Earth Sciences will find it useful. In order to present to the reader an up-to-date account of the various methods, this volume has been written in just over one year. The rather abbreviated style had to be adopted to accommodate the subject matter adequately in a book of reasonable size. It obviously would have been possible to include more details but, since the volume is intended to give a thorough outline only, the reader is referred to current publications for specialized information.

Throughout this book, emphasis has been laid upon the chemical and physical methods used, the significance of the results, and their interpretation in terms of investigating conventional geological problems. Many field geologists may never enter a laboratory equipped for dating, and therefore some detail is given to the principles of the various methods of analysis. While the mass spectrometer is perhaps the favourite tool of the geochronologist, other methods, such as neutron activation or X-ray fluorescence analysis, are often more suitable for certain types of analysis. Although current instrumentation may be complex, the whole framework of geochronology lies within the sphere of the geologist, without whom the useful interpretation of rock and mineral ages would not be possible.

The current trends in methods of analysis are directed towards improvement in both accuracy and precision of measurement. This applies particularly to the rubidium–strontium and lead–lead methods. In potassium–argon dating, it is already possible to date selected material less than one million years old. A major advance in the method is the description† of the relatively inexpensive A.E.I. MS 10 Mass Spectrometer for argon

---

* Gerling (1960), Hurley (1959), Knopf *et al.* (1931), Libby (1955), Rankama (1954, 1963), Russell and Farquhar (1960b), Starik (1961), Zeuner (1952, 1958). See References, commencing p. 229.

† Farrar *et al.* (1964). See References, p. 233.

analysis. This development makes it potentially possible for most Geological Institutes unable to support the expense of a full-scale Geological Age and Isotope Group to undertake some problems in potassium–argon dating.

Among the as yet relatively unexplored isotope pairs suitable for dating, that of rhenium–osmium should have a very interesting future, and will form an independent method against which strontium and lead isotope abundances may be compared. The samarium–neodymium pair ($^{147}Sm^{\alpha} \rightarrow {}^{143}Nd$) has not been used since the original suggestion by Wahl.* It would appear possible that the method could be used to date rare-earth rich minerals, particularly those found in alkaline rocks.

Apart from dating the geological age of rocks and minerals, variations in the isotopic compositions of strontium and lead offer an exciting field of research. It is perhaps by the use of this approach that some of the fundamental problems of geology will eventually be satisfactorily answered. A problem foremost in this field is the origin of igneous rocks and their relation to the earth's mantle. At present, attention is focused on basaltic rocks, although recent strontium and lead isotope studies suggest that the alkaline rocks may hold some important answers to this controversial problem.

I am indebted to Professor L. R. Wager of the Department of Geology and Mineralogy, Oxford, and to Dr. K. A. Davies of the Overseas Geological Surveys, London, through whose auspices a Department of Scientific and Technical Co-operation Research Fellowship has supported my period of research at Oxford. In addition, I am most grateful to Professor R. Farquhar and to Drs. K. Bell, S. Moorbath and N. J. Snelling for many useful discussions. Finally, to Mrs. J. McArdle my thanks for her kind assistance with the manuscript.

E. I. HAMILTON.

Oxford
March 1965

---

* Wahl (1942). See References, p. 245.

INTRODUCTION

THE COMPARATIVE GEOCHEMISTRY OF POTASSIUM RUBIDIUM, CALCIUM, ARGON, STRONTIUM URANIUM, THORIUM AND LEAD

L. H. Ahrens

Department of Geochemistry, University of Capetown, South Africa

The possibility of utilizing a given radioactive decay scheme for estimating the geological age of a mineral or rock rests ultimately on our ability to accurately estimate the parent (P) radioactive isotope and daughter (D) radiogenic isotope in a given mineral or rock.

The possibility of estimating P and D is naturally determined in part by our analytical prowess—the sensitivity and accuracy of the available analytical procedures. If the concentrations of both P and D are comparatively high, the analytical task is likely to be a simple one, but with decrease in concentration of P and D, the analytical problem becomes increasingly acute until, at very low concentrations, it may be quite impossible to undertake an age measurement.

The amount of P and D which may be found in a mineral or rock is controlled by several factors: the abundance of the element in the solar dust from which our planet originated, the isotopic composition of the parent and daughter elements in question, the half-life of the parent isotope and, finally, geochemical behaviour (in particular the extent to which the geochemistry of parent and daughter elements differ). Our main concern here will be comparative geochemical behaviour. Discussion will for the most part be confined to those radioactive elements which are commonly used for geochronological purposes, namely, K, Rb, U and Th, and their daughter elements; Ca and Ar (from K); Sr (from Rb) and Pb (from U and Th). Details of the respective decay schemes are given elsewhere in the appropriate chapters.

1*

Abundance data of these elements in chondrites (the commonest of all meteorites) and in some of the principal rock types (basaltic rocks, granitic rocks and shales), together with the crust, are as follows.

	K (%)	Rb (ppm)	Ca (%)	Sr (ppm)	U (ppm)	Th (ppm)	Pb (ppm)
Chondrites	0·085	2·5	1·40	11	0·014	0·04	$\begin{cases} 0\cdot15^a \\ 2\cdot5^b \end{cases}$
Basaltic rocks	0·7	30·0	6·7	470	0·6	2·5	5
Granitic rocks	3·3	150·0	1·6	290	4·8	17	20
Shale	2·7	130·0	2·2	300	3·7	12	20

[a] Common chondrites.   [b] Carbonaceous chondrites.

Only two of the elements under consideration are abundant, potassium and calcium. The remainder may be classed as trace elements. With the exception of lead, which shows both chalcophile and lithophile tendencies, and argon which is a gas, each element is distinctly lithophile.

I. Potassium and Rubidium

We will first consider the geochemical relationship between the two radioactive elements potassium and rubidium. Their chemistry and geochemistry are strikingly similar, which follows mainly from the fact that the radii of their respective cations, K^+ ($r = 1\cdot33$ Å) and Rb^+ ($r = 1\cdot45$ Å) are similar, and because the bonds they form with various ligands, including oxygen, are also of similar nature (ionic). Potassium is abundant and forms minerals, including the common varieties feldspar and mica; rubidium does not form minerals because of the ease with which Rb^+ is accepted into the structures of the potassium minerals where it may replace K^+. As one passes from the chondrites, as representative of the composition of the primitive dust from which our planet originated, through the common crustal igneous rock types, the ratio of potassium to rubidium remains more or less uniform and averages at about 250; that is to say, rubidium together with potassium enrich to about the same degree as one passes from basic to acid rocks. The spread of concentration (dispersion) of potassium tends, however, to be smaller than that of rubidium in the common igneous rocks basalt and granite. Thus, for example, the potassium concentration may vary from about 1% to 5% K (a factor of about five) in granite, rubidium varies from about 50 to about 400 ppm (a factor of almost ten).

A distinct enrichment of rubidium relative to potassium is usually apparent in roof granites and in particular in "small volume residuals" (aplite veins, some roof granites, pegmatites), reaching a maximum in lepidolite from pegmatite, where rubidium concentration averages about 1·5%, and K/Rb is about 6; this is the nearest approach to a rubidium mineral. Comparatively high rubidium concentrations (0·1–1·0%) may be found also in other pegmatite minerals, hydrothermal microcline including amazonite, zinnwaldite, some biotites and muscovites (particularly those that are lithium-rich) and the caesium mineral pollucite. The principal cause for the tendency of rubidium to enrich relative to potassium in minerals and rocks which may be described as "small volume residuals" is evidently due to the fact that Rb^+ is larger than K^+.

Within specific rock types, biotite and potash feldspar are the principal hosts for rubidium. However, the Rb concentration is not uniformly distributed and is usually preferentially accepted into the biotite structure; the concentration ratio Rb in biotite/Rb in K-feldspar is often in the region of two. The reason for this difference lies in the different structures of biotite and K-feldspar. The biotite structure is able to accommodate the relatively large Rb^+ (in comparatively high co-ordination) because it is more open than that of K-feldspar. An indication of the distribution of rubidium in biotite is provided by Figure D of Heier and Adams (1964, p. 342). The distribution is strongly positively skewed. Plagioclase contains distinctly lower rubidium concentrations than K-feldspar because of the difficulty of Rb^+ replacing relatively small Na^+ ($r = 0·97$ Å) and Ca^{2+} ($r = 1·01$ Å).

II. Calcium

Calcium is an abundant element and like potassium forms minerals of its own, notably the plagioclase of igneous rocks. In general, the geochemical behaviour of calcium contrasts quite sharply with that of potassium in igneous rocks—whereas the potassium concentration tends to rise with silica when passing from basic through intermediate to acid rocks, that of calcium tends to fall. Nevertheless, despite these opposing trends and the fact that the $K^+ \leftrightarrow Ca^{2+}$ replacement is not favoured, because of size difference, the high abundance of calcium together with the fact that its principal isotope (doubly "magic" $^{40}_{20}Ca$; $Z = N = 20$) is particularly abundant (97%) make it virtually impossible to distinguish radiogenic ^{40}Ca from non-radiogenic ^{40}Ca in potassium minerals. As a result, the decay $^{40}K \rightarrow ^{40}Ca$ is rarely usable. The situation is most favourable in some late-stage hydrothermal potassium minerals in pegmatite, notably lepidolite, where the total calcium concentration reaches a minimum and usually falls within the range 10–100 ppm (Ahrens, 1951).

If we turn to the sedimentary cycle we find one mineral, sylvite, in which the proportion of radiogenic calcium may be really high. Sylvite has been used for age determination purposes (Chapter 4).

III. Strontium

The geochemistry of this fairly rare alkaline earth metal is fairly closely linked to that of calcium mainly because doubly charged Sr^{2+} has a radius (1·18 Å) quite similar that that of Ca^{2+} and because both Ca^{2+} and Sr^{2+} form bonds with O^{2-} which are essentially ionic. Such behaviour may be contrasted with that of lead (below). Sr^{2+} may, however, also replace K^+ to some significant extent because the radii of these two cations do not differ greatly. The net effect of these two factors is that, when passing from basic to acid rocks, the strontium content does not fall sharply, as compared with calcium. Nevertheless, there is the general tendency for strontium to decrease in the sequence basic rocks→acid rocks→small volume residuals, and a minimum is reached in late-stage hydrothermal potassium minerals of pegmatites, notably lepidolite. It will be recalled that rubidium reaches its maximum in these minerals and accordingly the proportion of radiogenic strontium is at a maximum; in fact, strontium may be almost entirely radiogenic.

Within a given rock, strontium and rubidium do not vary sympathetically amongst the potassium and calcium (+sodium) minerals mica, K-feldspar and plagioclase. The position is such that in a spectrum of granite, one might find 20 ppm Sr in biotite, 100 ppm in the K-feldspar and distinctly more—perhaps 500 ppm—in the plagioclase. Rubidium will show the reverse tendency; perhaps 1000 ppm in the biotite, 500 ppm in the K-feldspar and less in the plagioclase. In two mica granites, muscovite tends to carry distinctly more strontium than biotite, whereas rubidium shows the reverse relationship.

The above geochemical relationships are only partly understood. For example, the rubidium relationship in biotite–K-feldspar–plagioclase follows from radius and crystal-structure grounds, but it is not clear why the rubidium concentration in biotite should be greater than in muscovite. Again, it is clear why the strontium content of plagioclase should be greater than that of K-feldspar, but it is more difficult to explain the very low strontium concentration in biotite, as compared with muscovite and K-feldspars. Because of the above relationships, biotite usually contains the highest proportion of radiogenic strontium of the rock-forming minerals and in this respect this mineral is particularly favourable for age measurements. Other rock forming minerals may, however, also be used as well as the rock as a whole (Chapter 5) provided that the rubidium and strontium concentrations are not highly unfavourable; that is, low rubidium and high strontium. This rules out the possibility, for example, of dating basic rock by the whole rock procedure because their rubidium concentration (average = 30 ppm) is too low and their strontium concentration (average = 470 ppm) is too high. Several granitic rocks have, however, been used for "whole rock" age measurements; whether or not such a rock can be used, however, depends on the rubidium and strontium contents. The available evidence indicates that whereas the rubidium concentration does not vary very greatly in granites (a factor of 10 or so is mentioned above), that of strontium does, and the possibility of using granitic rocks for "whole rock" age determinations depends more on the behaviour of strontium than on that of rubidium. An indication of the variation of strontium in granites is provided by the accom-

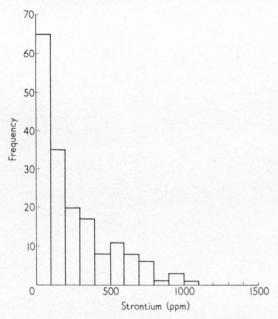

Frequency distribution diagram of strontium in 183 specimens of granite.

panying frequency distribution diagram. The data are mainly those of Turekian and Kulp (1956) and of Kaye *et al.* (1965). The distribution is strongly positively skewed and approximates lognormality. A high proportion of granites contain from <100–200 ppm, but quite high concentrations ranging up to ~1000 ppm may be found.

IV. Note on Argon

The geochemistry of the noble gas argon is quite distinct from that of all the other "geochronological" elements. Though compounds may be formed

by the comparatively inert noble gases, notably those between xenon and
fluorine, the atoms of argon in a potassium mineral will be in a free state,
more or less mechanically entrapped within the structure. The radius of the
argon atom is about 1·90 Å.

V. Uranium and Thorium

The characteristic oxidation state of thorium is four under almost any
conditions, whereas that of uranium may be four (igneous rocks) or six
(sedimentary environment); the higher oxidation state may also obtain in
secondary processes of igneous rocks. Quadrivalency of both uranium and
thorium will be assumed here in our discussion of the distribution of these
two elements in an igneous environment. The oxidation state of lead in an
igneous environment is normally two.

Thorium and uranium contents of minerals in igneous rocks (adapted from
(Adams, Osmond and Rogers, 1959)

Mineral	Thorium (ppm)	Uranium (ppm)
Major		
K-Feldspar	3–7	0·2–3·0
Plagioclase	0·5–3·0	0·2–5·0
Biotite	0·5–50	1–40
Hornblende	5–50	1–30
Pyroxene	2–25	0·01–40
Olivine	very low	
Accessory		
Allanite	500–5000	30–700
Apatite	20–150	5–150
Epidote	50–500	20–50
Monazite	25,000–200,000	500–300
Sphene	100–600	100–700
Zircon	100–2500	300–3000

For additional data, see reference cited above.

A particularly important feature about U^{4+} and Th^{4+} is their relatively large
size. The radii of these two cations (0·97 Å and 1·02 Å, respectively) together
with that of Ce^{4+} (0·94 Å) are far greater than the radii of any other naturally

occurring R^{4+} cations; Zr^{4+} (0·80 Å) is the next largest. Radii of highly charged abundant cations (Si^{4+}, Al^{3+}, Fe^{3+} and Ti^{4+}) are so very much smaller than ~ 1 Å that substitution of U^{4+} and Th^{4+} for Si^{4+}, Al^{3+}, Fe^{3+} and Ti^{4+} in their respective minerals is probably quite negligible. The only abundant cations with a radius magnitude of ~ 1 Å are Na^+ (0·98 Å) and Ca^{2+} (1·01 Å) but, because of the large charge differences, U^{4+} and Th^{4+} are not easily accepted into the plagioclase structure. The combined effect of the above two considerations is that uranium and thorium do not find a home in either ultrabasic or basic rocks, but tend instead to enrich as one proceeds to more acid rocks. The trend is therefore the same as that of Si, K and Rb. In fact, the trend of uranium and thorium may sometimes closely parallel that of potassium in a differentiated suite of igneous rocks. In general it seems that, within a given series of igneous rocks, the rate of increase of thorium concentration with respect to some index of differentiation may be greater than that of uranium, although the difference may not be large. The average chondritic and crustal ratio, Th/U, is approximately 4 and, although the same magnitude is found in most igneous rocks, there is often a general tendency for Th/U to increase somewhat in distinctly acidic rocks, for the reason given above. Th/U may fluctuate quite widely from specimen to specimen of a given rock type.

Because of the difficulty of acceptance of uranium and thorium in the common rock-forming minerals, these elements tend to concentrate in some of the accessory minerals, notably of granitic rocks. The table opposite (adapted from Adams, Osmond and Rogers, 1959) provides an indication of the concentration levels of uranium and thorium in different minerals from igneous rocks.

Although a major proportion of uranium and of thorium may be accommodated in the accessory and other minerals listed above, significant amounts may occur superficially along grain boundaries and within cracks.

Of the accessory minerals which are enriched in uranium and thorium, zircon has been particularly widely used for geochronological purposes; let us therefore look at the uranium and thorium distribution in this mineral more closely. Acceptance of uranium and thorium into the zircon structure follows from the fact that the radius of Zr^{4+} (0·80 Å) does not differ greatly from the radii of U^{4+} and Th^{4+}. Some uranium and thorium may, however, be present as uranothorite in the zircon.

The Th/U ratio averages at 0·4–0·5, a value distinctly lower than most rocks. Perhaps the preferential entry of uranium is due to the fact, in part at least, that the radius of U^{4+} is a little smaller than that of Th^{4+}. The concentration of both uranium and thorium varies widely in zircon (see table opposite). A fairly quantitative representation of the extent and nature of the variation of uranium and thorium in 70 zircons from granitic rocks is shown in the following figure, adapted from Ahrens (in preparation). It may be noted that the horizontal scale is logarithmic.

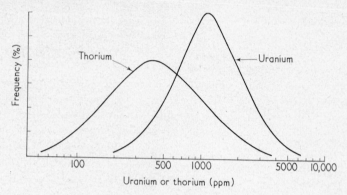

Distribution of uranium and of thorium in 70 accessory zircons from granites.

VI. Lead

The lead concentration tends to increase as one passes from basic through to acid igneous rocks and, in this respect, the trend shown by lead is similar to that of U, Th, K and Rb.

The stable oxidation state of lead in an igneous environment is two. The radius of Pb^{2+} is quite large (1·20 Å) and this ion can replace both Ca^{2+} (in plagioclase) and K^+ (in potash feldspar). In general, replacement of K^+ is greater than that of Ca^{2+}. The Pb^{II}—O bond, however, tends to have a significant degree of covalent character, which may account for the fact that, in a series of igneous rocks, the lead concentration tends to increase in the acid members. As lead shows chalcophile tendencies, some lead in an igneous rock may be present in the sulphide minerals.

Of particular significance for geochronology is the fact that Pb^{2+} is not readily accepted into the zircon structure, presumably because of large difference in size (cf. radii of Pb^{2+} and Zr^{4+}) and also a charge difference. As a result, the major proportion of lead in zircon and other highly accessory active minerals in igneous rocks tends to be radiogenic. In uranium and thorium minerals from pegmatites, lead is almost entirely radiogenic.

CONTENTS

CHAPTER 1

HISTORICAL INTRODUCTION

Before the application of radioactive dating methods, the ages of rock systems were ascertained by a study of the rates of sedimentation and denudation of geological material. After its discovery in 1896, radioactivity was soon applied to geological problems and from 1902 onwards rapid advances were made in this field. It soon became apparent that extremely accurate physical and chemical techniques were necessary to extract the maximum of value from radioactive measurements. These requirements were partially met during and after World War II with the development of atomic energy and its associated fields of study.

Xenophanes of Colophon (570–480 B.C.) was possibly the first to recognize that changes in fossil fauna reflected evidence of geological changes. The Greek historian Herodotos (484–424 B.C.) realized that the thick sediments deposited by the river Nile were a measure of the duration of prehistoric time. While scientists applied themselves to thoughts about geological time, religious estimates of the actual length of time were not based upon scientific principles. The Bible suggests that the earth is 6000 years old, but the derivation of this biblical age is rather vague. The Rabbinical chronology quotes a value of 3700 B.C. and the Roman Catholic of 5199 B.C. Other estimates, arrived at by the philosophical approaches of various Indian religious sects, and derived from a study of "cosmic-cycles" put the age at 4,320,000 years (Haber, 1959).

One of the first attempts to classify geological time in terms of periods was made by J. G. Lehmann in 1767. In the belief that parts of the original surface of the Earth still existed and that these did not contain organic life, they were called the Primitive Class of rocks. Fossiliferous rocks formed by secondary processes from the Primitive Class were called the Secondary Class, and a Third Class covered modern sediments with evidence of abundant organic life. Lehmann's results were elaborated upon by Werner (1780–1817), but the conclusions based upon a small area in northern Germany were incorrectly extrapolated on the assumption that these rocks were of world-wide occurrence. In France, Cuvier and Brongniart (1808) studied the minerals and organic remains in the neighbourhood of Paris and recognized ancient (extinct) and modern (living) fossils. These two types were used to sub-divide into groups the rocks in which they occurred. This approach was clarified by Lyell (1841), who classified European fossiliferous rocks into Periods and Groups in a form more or less the same as is in use at the present time. In a similar manner, Dana (1880) divided the major sequence of geologic time into Periods and these were further sub-divided into Epochs.

In a systematic measurement of geological time, Reade (1893) centred his arguments upon results obtained from the rate of denudation and accumulation of sediments. Reade estimated that a period of 95 million years covered the time since the Cambrian. Prior to this, Geikie (1822) had estimated the age of the earth as 100–600 million years. During the late eighteenth century, geologists and physicists disagreed as to the length of geological time. The physicists placed a limit much less than that required by geologists. Lord Kelvin estimated that the time required for the earth to have cooled sufficiently to support organic life could not be greater than 100 million years and was probably nearer 25 million years. If the earth was a fluid mass, the theories of the rotation of a spherical mass required a certain amount of global flattening which increased with time. If the earth had solidified 100 million years ago, the present shape would be concurrent with such a hypothesis. As the age of the sun cannot be greater than that of the earth, evidence from the mutual gravitational contraction of the solar material placed a possible limit of 100 million years. That the sun could have illuminated the face of the earth 500 million years ago was considered quite impossible. The most important factor governing Kelvin's estimates was the assumption of a cooling earth; the development of an internal source of heat by chemical reactions was considered only a minor factor that would not prolong the cooling period by very much. However, Kelvin did qualify his finding in that this hypothesis would only be true if an internal source of heat was not present, and at that time none was known. This estimate became meaningless when the heat production from radioactive elements was discovered. In 1893, Williams introduced the term *geochronology* in which the duration of time represented by the Eocene was taken as a standard unit, a geochrone. The subdivisions of geological time based upon the thickness of sediments in terms of geochrones resulted in Williams's Time Scale of:

Recent	→ Eocene	3 geochrones
Cretaceous	→ Triassic	9 geochrones
Carboniferous	→ Cambrian	45 geochrones

Using generalized values, Goodchild (1897) calculated the length of time necessary for the deposition of post-Cambrian sediments by the rate of sedimentation of the most abundant sedimentary rocks. The values used were:

1 foot of sandstone	accumulated in	1500 years
1 foot of shale	accumulated in	3000 years
1 foot of limestone	accumulated in	25,000 years

The total estimated age of 704 million years is in good agreement with modern values based upon radioactive dating.

A totally different method of dating by the sodium content of ocean water was made by Joly (1899) and Lane (1929). The age is normally calculated by dividing the total sodium content of the oceans by the amount of sodium entering them through river transport. The minimum age calculated by this method is subject to many errors such as salt entrapped in sedimentary rocks and the amount of ocean salt contributed from volcanic sources. A recent appreciation of the sodium cycle in terms of the age of oceans and sediments has been given by Livingstone (1963).

Biologists were also active in contributing to the age of the earth by independent methods. The main approach comes from Lamarck in his work on the adaptation of organic life to environment, and Darwin of natural selection. The evolution of mammalian phyla with time was described by Matthew (1914).

The discovery of radioactivity marked the beginning of a new era in geochronology. Before a radioactive atom decays, its chemical characteristics are the same as those of other atoms isotopic with it, but when a radioactive atom does decay it emits energetic radiation that can be detected. It is through the latter that radioactivity was first realized. The application of radioactivity to the dating of rocks and minerals depends upon four main requisites.

(1) A parent atom A by a random process becomes radioactive.

(2) The parent atom A, by radioactive disintegration, is transformed (decays) to a daughter atom B.

(3) The rate at which the reaction A→B proceeds is a constant and must be accurately known.

(4) The system remains closed with regard to parent and daughter after the geological event being studied, e.g. time of crystallization of an igneous rock.

If the present concentration of A and the amount of daughter B can be determined by chemical or physical methods then, knowing the rate at which B has been formed, an age can be estimated. Holmes has likened the geological application of radioactivity to a clock wound up at the time of crystallization of a rock or mineral. Consider a granite magma in which it is assumed, for present purposes, that all the elements are homogeneously distributed throughout the magma. After crystallization, the various elements have been segregated into the constituent minerals. If a radioactive element is concentrated in a particular mineral then at the time of crystallization the mineral will contain only the pure component A. With the passage of time the daughter B is continuously being produced and accumulates within the mineral. The rate of change A→B is known, and this is analogous to the gradual running down of the radioactive clock; the proportion of B present indicates the time that has passed.

It is necessary that the rate of change A→B is not very long in relation to the age of the sample, otherwise the amount of B produced will be too small to be detected even by refined physical and chemical techniques.

The radioactive elements found in nature are given in Table 1 together with their isotopic abundances and their decay rates in terms of half-lives.

TABLE 1. Naturally occurring radioactive elements

Parent	% Abundance	Daughter and mode of decay	Half-life[a] (years)
K^{40}	0·0119	e ^{40}Ar β^- ^{40}Ca[b]	$1·33 \times 10^9$
Rb^{87}	27·85	β^- ^{87}Sr	$5·0 \times 10^{10}$
In^{115}	95·77	β^- ^{115}Sn	$6·0 \times 10^{14}$
La^{138}	0·089	e ^{138}Ba β^- ^{135}Ce[b]	$\sim 7·0 \times 10^{10}$
Sm^{147}	15·09	α ^{143}Nd	$1·25 \times 10^{16}$
Lu^{176}	2·59	e ^{176}Tb β^- ^{176}Hf[b]	$2·4 \times 10^{10}$
Re^{187}	62·93	β^- ^{187}Os	$\sim 5 \times 10^{10}$
Th^{232}	100·00	Complex ^{208}Pb	$1·39 \times 10^{10}$
U^{235}	0·72	Complex ^{207}Pb	$7·1 \times 10^8$
U^{238}	99·27	Complex ^{206}Pb	$4·5 \times 10^9$

[a] See Chapter 2.
[b] Branching mode of decay.

Although the radioactive decay series of uranium and thorium (Table 33) are now well known, it was in 1902 that Rutherford and Soddy first investigated the radioactive properties of thorium. By simple chemical separations, they were able to show that radioactive species could be separated from a thorium salt such that the separated components had distinct chemical and radioactive properties. They also showed that the normal or constant radioactivity possessed by thorium is an equilibrium value, where the rate of increase of radioactivity due to the production of fresh active material is balanced by the rate of decay in radioactivity of that already formed. The detection and identification of radioactive properties were achieved by studying the darkening of a photographic plate, the "leakage" rate of a simple electroscope and the penetration properties of radioactivity through different thicknesses of metal foil. These physical determinations were coupled with separations by simple chemical techniques.

Strutt in 1905 determined the amount of uranium, radium, thorium and helium in a variety of minerals. The results showed that the amount of radium present was proportional to the amount of uranium, and helium was present only in uranium or thorium bearing minerals. Boltwood (1907) studied the end disintegration products of uranium and concluded that the ratio Pb/U in primary uranium minerals was dependent upon geological age, as shown in Table 2. Lead and helium were not regarded as disintegration products of thorium and it was not until 1914 that Soddy and Hyman showed that thorium, like uranium, resulted in the production of both lead and helium.

TABLE 2. Variations of Pb/U ratios with geological age

Geological Period	Pb/U	Age (years $\times 10^6$)
Carboniferous	0·041	340
Devonian	0·045	370
Silurian-Ordovician	0·053	430
Pre-Cambrian		
Sweden	$\begin{cases} 0·125 \\ 0·155 \end{cases}$	1025 1270
U.S.A.	$\begin{cases} 0·160 \\ 0·175 \end{cases}$	1310 1435
Ceylon	0·200	1640

In a similar manner, Strutt (1908, 1909) observed that the He/U ratio was related to geological age and that, while high values were not found in young deposits, low values were found in old deposits, indicating helium loss. The helium ratio used by Strutt accounts for the total helium production from both uranium and thorium upon the basis that 1 g Th is equivalent to 0·203 g U_3O_8. Apart from uranium minerals, a study of a resistant mineral such as zircon showed a similar ratio increase with time but, still accepting helium loss, the ratio indicated minimum values. Helium ratios from zircons are given in Table 3. Chemical methods current during this period were unable to measure U, Th, Pb, He in normal accessory zircon, and the measurements are restricted to pegmatite occurrences.

Holmes (1913) applied the association of lead and uranium to the measurement of geologic time. Using a decay rate of 1 mole of lead replacing 1 mole of uranium in 8200 million years, minerals of the same age should have a constant Pb/U ratio provided that:

(a) no lead was present in the original material;
(b) lead was not produced by any other radioactive process;
(c) there was no addition or subtraction of uranium or lead by any external agents.

The uranium content was determined by the Ra emanation method, and that of lead by classical sulphate precipitation or a colorimetric method using the colour of finely divided lead sulphide.

TABLE 3. Helium ratios of zircons from various localities[a]

Geological age	Locality	$He \times 10^2$	$U \times 10^2$	$Th \times 10^2$	$0 \cdot 27 \, Th \times 10^2$	$\dfrac{He}{U + 0 \cdot 27 \, T}$
Recent	Vesuvius, Italy	0·4	32·2	—	—	0·012
Pleistocene	Eifel, Germany	1·14	10·77	none	—	0·110
Miocene	Auvergne, France	2·12	3·15	none	—	0·670
Jurassic	Tasmania	4·34	0·97	none	—	4·470
Probably late Carboniferous	Green River, N. Carolina	255·0	10·94	26·5	7·16	14·070
Probably late Upper Pre-Cambrian	Ceylon gem gravels	283·0	8·56	3·52	0·95	29·77
Middle or Lower Pre-Cambrian	Renfrew Co., Ontario	114·0	1·55	0·81	0·27	64·04

[a] Strutt (1909); see Knopf, Schuchert, Kovarik, Holmes and Brown (1931).

The possibility of determining geological age by the examination of pleochroic haloes surrounding radioactive minerals in micas was investigated by Joly (1907, 1923). The time required to produce the amount of darkening was compared to the amount of artificial darkening produced by bombarding a mica with α-particles from a radium source. Although the presence of thorium could be detected by measurement of the constituent rings forming a given halo, chemical methods were not sufficiently advanced to determine trace amounts of uranium and thorium in such accessory minerals. The method is subject to many difficulties, such as the darkening reaching saturation level with time and the fading or loss of the halo with increase in temperature associated with igneous activity.

With the recognition of three radioactive series (^{238}U, ^{235}U, ^{232}Th) and the production of lead as an end product, attention was focused upon the determination of the atomic weight of lead from various sources. For this purpose it was necessary to distinguish between the following types of lead.

(i) *Primeval lead*. Lead present in the earth when it was formed is assumed to have had an identical isotopic composition.

(ii) *Common lead.* This is not formed from radioactive processes and comprises the main lead ores such as galena (PbS) and cerussite (PbCO$_3$) together with lead at trace levels incorporated in the main rock-forming minerals, particularly potash feldspar. Common lead consists of primeval lead together with traces of radiogenic lead incorporated during magmatic and sedimentary processes.

(iii) *Radiogenic lead.* Lead derived as a stable end product of the radioactive decay of ^{238}U, ^{235}U and ^{232}Th. Radiogenic leads are derived through the emission of α-particles (^{4}He$^+$) resulting in a decrease in atomic weight of 4 mass units with the emission of each α-particle. The atomic weight of ordinary lead was determined to be 207·2, and in 1914 Soddy and Hyman suggested that minerals containing uranium, but no thorium, should have lead of smaller atomic weight than ordinary lead, while that from thorium rich minerals should be greater. Soddy and Hyman showed this to be correct for lead separated from thorium minerals, while Richards and Lembert (1914) independently showed the same for lead from uranium rich minerals. The atomic mass changes for lead derived from radioactive process are shown in the following sequences.

$$^{238}\text{U} \rightarrow {}^{206}\text{Pb} + 8 \ (^4\text{He}^+) \text{ mass depletion of } 32$$

$$^{235}\text{U} \rightarrow {}^{207}\text{Pb} + 7 \ (^4\text{He}^+) \text{ mass depletion of } 28$$

$$^{232}\text{Th} \rightarrow {}^{208}\text{Pb} + 6 \ (^4\text{He}^+) \text{ mass depletion of } 24$$

Before the rapidly expanding field of dating by means of radioactive measurements was firmly established, the estimates of age based upon sedimentation were summarized by Barrell (1917). He described the non-uniformity of erosion in terms of accelerated periods of uplift followed by denudation. These sporadic uplifts must give rise to a sedimentation rate much greater than the mean of geological time. Although a unit area of rock subject to erosion produces a certain amount of sediment, the subsequent deposition of this material in a secondary unit of sediment does not equal in weight or volume the original source material, because a portion of the sediment is carried away through the agents of transport. The total thickness of a pile of sediments is often less than the true value because of major and minor unconformities. Although some stratigraphic breaks are recognized by the fossil record, this is not always the case. In contrast, radioactive dating is independent of such sources of error, and the radiometric dating of rocks has become a reality. Igneous rocks could be dated and the age patterns of the Pre-Cambrian studied. The time scale of Barrell is given in Table 4.

In 1927 a major analytical advance in geochronology was made by Aston. For the first time the mass spectrograph was used to study the isotopic composition of lead. With lead tetramethyl as a source of lead and using photographic plate recording, he observed lead isotopes at masses 206–207–208.

The spectrum was complicated by $HgCH_3$ derived from mercury used in the vacuum system, but the possibility of masses at 203, 204, 205 was noted.

Schuler and Jones (1932) proved the presence of a lead isotope of non-radiogenic origin at mass 204. Lead-204 could now be used to determine the amount of primary lead present in minerals. A systematic determination of the isotopic composition of lead by mass spectrometry was carried out by Nier (1938) and Nier, Thompson and Murphy (1941). While the mean atomic weight of lead varied but little, the independent abundance of ^{206}Pb, ^{207}Pb, ^{208}Pb (referred to $^{204}Pb = 1\cdot000$) showed considerable variations.

TABLE 4. Barrell's geological time scale

Period	Minimum	Maximum (years $\times 10^6$)
Recent	1	1·5
Cenozoic	55	65
Mesozoic	135	180
Palaeozoic	360	540

After Barrell (1917)

Holmes (1913, 1929, 1931) argued persuasively that the new methods of estimating ages were potentially far more reliable than those of classical geology. During and after the 1939–45 war, analytical methods were greatly improved, in particular those of mass spectrometry, and the determination of trace amounts of elements. Two new methods of dating were evolved during this period, those of potassium–argon ($^{40}K \rightarrow ^{40}Ar$, ^{40}Ca) and rubidium–strontium ($^{87}Rb \rightarrow ^{87}Sr$). It is the latter two methods and those involving studies of lead isotopes that constitute the most widely used dating methods in current use. In the present state of geochronology, the distribution of radiogenic isotopes has assumed great importance in the understanding of geological events and petrological problems.

CHAPTER 2

GENERAL METHODS

I. Radioactivity

When a radioactive nuclide undergoes a process of nuclear disintegration, the rate of radioactive decay is proportional to the number N of reactant nuclei present, i.e.

$$\frac{-\mathrm{d}N}{\mathrm{d}t} = \lambda N \tag{1}$$

where N is the number of radioactive atoms of original parent remaining after a period of time t, and λ is a decay constant representing the probability that an atom will disintegrate in unit time.

The relation giving explicitly the number of radioactive parent atoms remaining after time t is obtained by integration of equation (1).

$$N = N_0 e^{-\lambda t} \tag{2}$$

or

$$N_0 = N e^{\lambda t} \tag{3}$$

where N_0 is the number of atoms at a time "t".

In practice, changes due to the decay rate are expressed in terms of half-lives. By definition, a half-life is the time required for an amount, or the radioactivity of a radioactive substance, to be reduced to one-half of its initial value. The condition for the half-life is then $N/N_0 = \frac{1}{2}$. Equation (3) gives:

$$N/N_0 = e^{-\lambda t} \tag{4}$$

substituting $\frac{1}{2}$ for N/N_0

$$\frac{1}{2} = e^{-\lambda t} \tag{5}$$

and T (= half-life) for t:

$$2 = e^{\lambda T} \tag{6}$$

$$T = \log_e 2/\lambda = 0 \cdot 6931/\lambda \tag{7}$$

The half-life is inversely related to the decay constant and is therefore independent of the initial amount of radioactive material.

The radiation emitted from a sample can be detected by many techniques but in general these methods yield a number A that is proportional to $-dN/dt$. Therefore, the general form for equation (2) is

$$A = A_0 e^{-\lambda t} \tag{8}$$

where A is the activity of disintegrating material at a time t and A_0 is the initial radioactivity.

In radiometric dating it is convenient to measure time backwards from the present. Consequently the basic equation of geochronology is

$$\frac{D}{P} = e^{\lambda t} - 1 \tag{9}$$

where D is the present number of atoms of daughter nuclide in a sample of rock or mineral formed since a time t (A in equation (9)); P is the present number of atoms of parent nuclide in a sample of rock or mineral (A_0 in equation (9)); λ is the decay constant of the parent nuclide and t is the time measured from the present. The apparent age (t) of the sample is given by the equation:

$$t = \frac{1}{\lambda} \log_e \left(1 + \frac{D}{P} \right) \tag{10}$$

In some cases the apparent age may equal the true age, provided that (a) there has been no net migration of parent or daughter nuclides into or out of the rock or mineral during its lifetime—i.e. a closed system; (b) the rate of decay Parent→Daughter is known with sufficient accuracy; (c) the sample is representative of the system being studied; (d) there has been no addition of extra amounts of parent or daughter nuclide through contamination during the chemical and physical procedures used; (e) the accuracy of the analytical method is known.

II. Stable Isotope Dilution Method

The measurement of daughter and parent nuclides is best determined by the method of stable isotope dilution analysis. An accurately known quantity of tracer consisting of the element having an isotopic composition very different from that of the natural element, is mixed with a solution of the sample. The element to be determined is then chemically separated and the isotopic composition of the mixture (tracer plus natural element) is measured by mass spectrometry. The quantity of natural element is determined by measuring the change in the tracer caused by the presence of the element having a normal isotopic composition.

A typical sequence of events used in determining the concentration of an element in a rock or mineral is as follows. A solution containing the tracer, generally prepared by electromagnetic separation, is dissolved in a suitable reagent. It is not generally convenient to measure the exact concentration of tracer by gravimetric analysis as it is generally expensive and only a few milligrams are diluted to known volume. Instead, the concentration is determined by calibrating the tracer with a solution containing an accurately known amount of the element having a normal isotopic composition. The chemical purity of the tracer is not important as it is to be calibrated, but the anion and cation impurities in the standard must be known accurately. In the case of gaseous samples, identical steps are carried out using conventional gas analysis techniques.

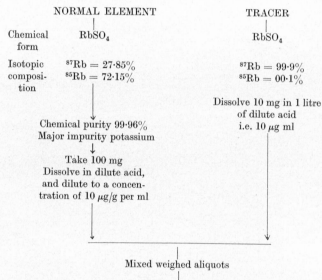

NORMAL ELEMENT

TRACER

Chemical form $RbSO_4$

$RbSO_4$

Isotopic composition $^{87}Rb = 27 \cdot 85\%$ $^{85}Rb = 72 \cdot 15\%$

$^{87}Rb = 99 \cdot 9\%$ $^{85}Rb = 00 \cdot 1\%$

Dissolve 10 mg in 1 litre of dilute acid i.e. 10 μg ml

Chemical purity 99·96% Major impurity potassium

Take 100 mg Dissolve in dilute acid, and dilute to a concentration of 10 μg/g per ml

Mixed weighed aliquots

Determine isotopic composition of mixtures and, by isotope dilution analysis, calculate concentration of ^{87}Rb in tracer solution.

FIG. 1. Procedure for the calibration of rubidium tracer.

To a solution of a sample, a known amount of the tracer is added. Once sample and tracer have been completely mixed, a quantitative separation of the element is not necessary as isotope ratios are measured, not absolute quantities. The solution is then processed to remove the element required in a pure form. If the subsequent mass spectrometry is to be carried out by solid state emission, any impurities present will generally lead to poor ion emission. The separated element consists of a mixture of normal and tracer isotopes. The change in isotopic composition of the element to be determined, caused by the addition of the tracer isotope, is determined by mass spectrometry.

A typical isotope dilution calibration for rubidium is given in Fig. 1.

Examples (a) and (b) which follow make use of the stable isotope dilution method of analysis.

(a) *Analysis of an element containing two isotopes one of which is present in an overwhelming amount* (He, Li, V, La)

Isotopic composition of sample $X = X^1$ isotopic abundance $>99\%$
X^2 isotopic abundance $<1\%$

Isotopic composition of tracer $X_t = X^1_t$ isotopic abundance $<1\%$
X^2_t isotopic abundance $>99\%$

In the general case it can be assumed that the sample consists only of X^1 and the tracer of X^2_t. A one gram sample of a rock or mineral is dissolved in mixed acid and $1\,\mu g$ of the tracer X^2_t is added. The mixture is chemically

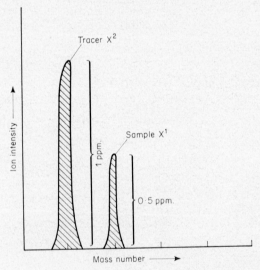

Fig. 2. Diagram representing an isotope dilution analysis of element X separated from a sample.

processed to separate out the element X in a pure form and its isotopic composition determined by mass spectrometry. The ratio of sample to tracer $= 1:2$, but the peak height of the tracer $= 1$ ppm (i.e. $1\,\mu g$ per g sample $=$ concn of 1 ppm); therefore the sample contains 0·5 ppm of X. This type of analysis is illustrated in Fig. 2.

(b) *Analysis of an element containing more than one isotope in significant amounts in both sample and tracer*

When an element contains more than two isotopes, the normal tracer consists of a mixture of isotopes, and it is necessary to solve two sets of simultaneous equations. This may be illustrated by strontium, which consists of the following isotopes

^{88}Sr	82·56%
^{87}Sr	7·02%
^{86}Sr	9·87%
^{84}Sr	0·55%

The $^{88}Sr/^{86}Sr$ ratio, which is considered to be invariant in nature, can be used to determine the total strontium content of a sample by isotope dilution analysis. Strontium-87 is variable, as it can consist of normal ^{87}Sr + radiogenic ^{87}Sr from the decay of ^{87}Rb.

The relative amount of ^{84}Sr is too small for accurate comparative measurement. It is common practice to use ^{86}Sr as a tracer, even though it is not normally produced in a pure form (i.e. ∼80% ^{86}Sr).

The isotopic ratios (R) measured for normal, tracer and mixture consist of

$$R_1 = \frac{(^{86}Sr)}{(^{88}Sr)} \quad \text{(normal)}$$

$$R_2 = \frac{(^{86}Sr)}{(^{88}Sr)} \quad \text{(tracer)}$$

$$R_3 = \frac{(^{86}Sr)}{(^{88}Sr)} \quad \text{(mixture)}$$

The mass of Sr in the sample is unknown and this is solved from R_1, R_2, R_3 and the known mass of strontium in the tracer.

$$\text{Mass of Sr in sample} = \frac{(R_1+1)(1-R_3/R_2)}{(R_3-R_1)(1+1/R_2)} \times \text{mass of Sr in tracer}$$

A general equation for solving the problem has been given by Webster (1958).

$$x = y \frac{B-C}{C-A} \frac{M_A}{M_B}$$

x = the weight of normal element
y = weight of tracer added
A = the ratio of isotopes in the normal element
B = the ratio of isotopes in the tracer element
C = the ratio of isotopes in the mixture

M_A = the chemical atomic weight of normal element

M_B = the chemical atomic weight of tracer

M_A, M_B can be calculated from the known isotopic abundances.

When the element has four or more isotopes, it is possible to make use of a third independent ratio to calculate the mass discrimination that arises during the mass spectrometry in any one analysis. This approach has been used by several workers in the determination of strontium. The calculation of optimum quantities of tracer to be added to different amounts of the sample to be measured becomes involved, but Crouch and Webster (1963) have described the required conditions.

In the mass spectrometric measurement, sources of isobaric interference must be minimized. For example, rubidium, consisting of ^{87}Rb–27% ^{85}Rb–75%, is very easily ionized and, although it starts emitting at a lower filament temperature, its emission range covers that at which strontium ions appear. In such an instance, ^{87}Rb causes isobaric interference when determining the height of the ^{87}Sr peak. The problem can be partly overcome by

TABLE 5. Scope of isotope dilution method of analysis

Elements suitable for I.D. analysis				Mono-isotopic elements not suitable for I.D. analysis
H	V	Pb	Gd	F
He	Cr	Ag	Dy	Na
Li	Fe	Cd	Er	Al
Be[a]	Ni	In	Tb	P
B	Cu	Sn	Lu	Sc
C	Zn	Sb	Hf	Mn
N	Ga	Te	W	Co
O	Ge	I[a]	Re	As
Ne	Se	Xe	Os	Y
Mg	Br	Cs[a]	Ir	Nb
Si	Kr	Ba	Pt	Rh
S	Rb	La	Hg	Pr
Cl	Sr	Ce	Tl	Tb
A	Zr	Nd	Pb	Ho
Ca	Mo	Pm[a]	Th	Tm
Ti	Tc[a]	Sm	U	Ta
K	Ru	Eu	Pu[a]	Au
				Bi

[a] Tracer prepared in weighable amounts from long-lived isotopes produced in a nuclear reactor.

determining the ^{87}Rb/^{85}Rb ratio before strontium appears in the spectrum. During the emission of strontium, the increment of ^{87}Rb added to the ^{87}Sr peak can be calculated by measuring the ^{85}Rb peak present in the strontium spectrum. As the ^{87}Rb/^{85}Rb ratio is known, the ^{87}Sr peak can be multiplied by a suitable correction ratio.

The isotope dilution method can be applied to about 80% of the elements, and those that cannot be analysed by this method or are more suitably determined by other techniques are given in Table 5. The sensitivity varies with the elements, but values range from a few micrograms down to 10^{-13} g or better. The sensitivity and precision of isotope dilution analysis for uranium and thorium are given in Table 6. An accuracy of a few per cent should be attainable even at very low concentrations.

TABLE 6. Examples of the sensitivity and precision of isotope dilution analysis

Element	Sample	Results (mean+S.D.)	
Sensitivity			
U	Ammonium nitrate reagent	$0 \cdot 000075 \pm 0 \cdot 000004$ ppm[a]	
	Stone meteorite (Modoc)	$0 \cdot 0105$ ppm	
	Perthite feldspar	$0 \cdot 22 \pm 0 \cdot 03$ ppm	
Th	Perthite feldspar	$0 \cdot 410 \pm 0 \cdot 008$ ppm[b]	
Precision			
Uranium[c]	Sea-water	$3 \cdot 16$	
		$3 \cdot 26$	
		$3 \cdot 28$	$3 \cdot 26 \pm 0 \cdot 06 \, \mu g/litre$
		$3 \cdot 29$	
		$3 \cdot 32$	

[a] Tilton (1951).
[b] Tilton, Aldrich and Inghram (1954).
[c] Rona, Gilpatrick and Jeffery (1956).

III. Mass Spectrometry

A mass spectrometer is an instrument that produces a beam (ion current) of ions (emitted from a sample), and separates them according to the mass-to-charge ratio. In general terms, a mass spectrometer consists of the following components.

(1) A sample system by which the sample is introduced into the mass spectrometer.

(2) An ion source to produce ions characteristic of the sample.

(3) A system of slits through which the ion beam passes, is accelerated through an electrical field, and collimated. The ion beam can be focused for (a) *direction* (focused for a range of different initial directions, when all the ions in the beam are moving at the same speed), (b) *velocity* (ions travelling with a range of different speeds but all moving in the same direction), (c) *double focusing* (focusing of ions of varying initial speed and direction).

(4) An analyser section (magnet) in which the beam is resolved.

(5) A final slit system through which the separated ion beam passes.

(6) A detector system by which the resolved ion beam is rendered detectable.

(7) A system for recording the relative intensities of the separated ions, e.g. galvanometer, chart recorder.

To prevent positive ion scattering and ion energy changes as a result of intermolecular collisions, the total path of the ion beam must be in a region of low gas pressure. Under such conditions the rate "ion flow" is essentially limited by the frequency of collision with the internal walls of the mass spectrometer. This necessity has led to the development of high vacuum techniques in modern spectrometry. Most mass spectrometers used in geochronology are of conventional design and a schematic diagram of a single-direction focusing mass spectrometer is given in Fig. 3(a).

Figure 3(b, c) illustrates two commercial mass spectromers manufactured by the Associated Electrical Industries, England. The MS2 machines can be used for the analysis of solids and gases over the total mass range, although in the illustration only the gas inlet system is visible. Compared to this the MS10 is a relatively small and inexpensive spectrometer and the cost is approximately one-tenth that of the MS2. Its range is limited to masses 2–100 but is certainly capable of being used for potassium–argon dating and should have an interesting future in measuring the amount and isotopic composition of gases present in geological samples. While the illustrations may be of more interest to specialists in the field of mass spectrometry, they at least show to the geologist the range of size and shape of mass spectrometers, although even larger and smaller types are produced for special projects.

The term mass spectrograph is conventionally used to describe an instrument in which the final ion beam is recorded on a photographic plate, while the term mass spectrometer is used when the ion beam is recorded by electrical means.

In an ideal mass spectrometer the intensity of the ion current produced by a particular isotope is proportional to the concentration of that isotope. Another requirement of the instrument is that it is able to resolve one ion beam of mass A from an ion beam of (approximately the same) mass $A+1$. Mass spectrometers of high resolution are essential in the analysis of the complex spectrum of organic material.

The literature describing the design of mass spectrometers is extremely voluminous, and many detailed variations have been developed for the different components. It is beyond the scope of this book to describe the different types and the reader is referred to standard works on this subject (Nier, 1947; Inghram and Hayden, 1954; Aston, 1942; Barnard, 1956; Duckworth, 1958; Ewald and Hintenberger, 1953; Hintenberger, 1957; Beynon, 1960).

As the final data for solving geological age equations depend upon the quality of the mass spectrometer results, some detail is given here to a description of the more important components.

A. ION SOURCES

1. Gas source

A commonly used gas source is described by Nier (1940, 1947) and Reynolds (1956). The sample of gas is introduced into the mass spectrometer through a controlled leak (Inghram, 1948; Barnard, 1953) and then into the ionization chamber, where it is subjected to bombardment by electrons from an incandescent filament. Electron bombardment removes one or more electrons from a neutral atom leaving a positive particle. In relation to the electron, the positive particle is massive and moves with a lower velocity. For this reason, the region through which the particle passes must be under very low gas pressure. As a result of passing through a region of high electrical potential gradient, it accumulates sufficient energy to become detectable. The ionizing beam of electrons is carefully controlled since in single-focusing machines the focusing properties of the magnet are realized only for ions of the same charge : mass ratio and of the same velocity. There is a minimum threshold energy below which ionization does not occur; the production of an excessively intensive electron beam results in a large spread of velocity and leads to poor resolution.

In addition to pure gases, the gas type of source is also used for volatile liquids and compounds of relatively high vapour pressure, such as lead tetramethyl and silicon tetrafluoride, for the determination of the isotopic composition of lead and silicon.

2. Solid sources

(a) *Filament surface ionization*

Surface ionization techniques were first used by Gehrke and Reichenheim (1906, 1907). The application of this technique to modern mass spectrometry has been described by Inghram and Hayden (1954) and Palmer (1956).

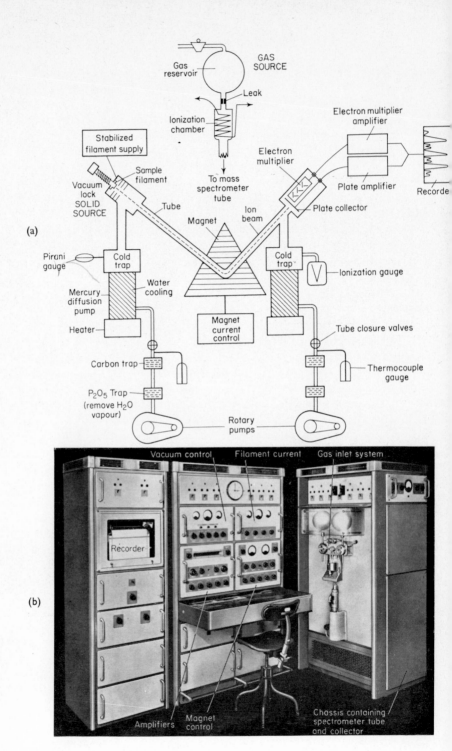

(a)

Gas reservoir

GAS SOURCE

Leak

Ionization chamber

Stabilized filament supply

Sample filament

Vacuum lock
SOLID SOURCE

Tube

To mass spectrometer tube

Electron multiplier amplifier

Electron multiplier

Plate amplifier

Recorde[r]

Magnet

Ion beam

Plate collector

Pirani gauge

Cold trap

Water cooling

Cold trap

Ionization gauge

Mercury diffusion pump

Heater

Magnet current control

Tube closure valves

Carbon trap

Thermocouple gauge

P_2O_5 Trap (remove H_2O vapour)

Rotary pumps

(b)

Vacuum control   Filament current   Gas inlet system

Recorder

Amplifiers   Magnet control

Chassis containing spectrometer tube and collector

(c)

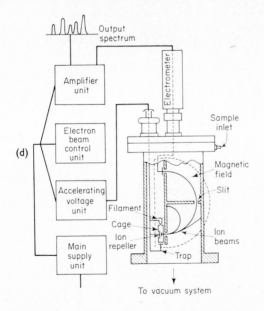

(d)

Fig. 3. (a) Diagram of mass spectrometer flow system. (b) Commercial Associated Electrical Industries MS2 Mass Spectrometer (photographs kindly supplied by A.E.I.). (c) Associated Electrical Industries MS10 Mass Spectrometer. (d) Diagrammatic layout of MS10 mass spectrometer.

The rate of production of positive to neutral particles produced by evaporation from a heated filament surface is given by the equation

$$\frac{n^+}{n^0} \propto \exp\left[e(w-1)/kt\right] \tag{11}$$

where n^+/n^0 is ratio of neutral to positive particles produced, e is the electron charge, w is the work function of filament surface, t is the absolute temperature of the filament and k = Boltzmann's constant. The equation expresses the probability that a particle evaporating from the filament surface will be emitted as a positive ion.

From equation (11) it is apparent that the filament material should have a high work function and low ionization potential. The work function of a metal is defined as the minimum additional energy which would enable an electron to escape from the metal, while the ionization potential in electron volts (eV) is the least energy necessary to remove one electron from a free unexcited neutral atom. For a sample of a refractory nature it is also necessary that the filament material has a high melting point. Although the work function, ionization potential and melting point in present literature vary a little, some values are given in Table 7.

TABLE 7. Work functions for filament materials used in solid sources

Element	Work function (thermionic) (V)	Work function (photoelectric) (V)	Ionization potential (eV)	Melting point (°C)
C	3·34	4·81	—	3500
Ta	4·12	4·05	—	3000
W	4·37	4·15	7·98	3380
Rh	4·80	4·57	7·7	3170
Pt	5·32	6·30	8·96	1769
Th	3·38	3·5	—	1700

The main variable in equation (11) is the work function. The production of stable carbides, oxides, or nitrides while the filament is heated produces different values for the work function. A tungsten surface with a monoatomic layer of oxygen is stable up to 1800°C, while the work function has increased to 9·2 V (Longmuir and Kingdon, 1925). Although tantalum and tungsten are the main materials used for filaments, rhenium (Sharkey, Robinson and Friedel, 1959) has become increasingly popular. The metal is ductile at all temperatures and the carbides are unstable, an important point when analysing hydrocarbons. Sintered tantalum filaments have also been used with some success (Goris, 1962; Hamilton, 1963). When a mixture of $Ta + Ta_2O_5$ powders is heated in the absence of oxygen, the oxide is reduced to the metal and produces a very large porous filament surface area; it is essential that only sufficient Ta_2O_5 is used so that the Ta powder is only just sintered to a coherent mass. For efficient ionization the sample is thinly spread over the surface of the filament. In many cases the separated element is often contaminated with other elements, resulting in a rather thick layer being placed on the filament. The sintered surface with a very large surface area is advantageous in this case, as a thick deposit can lead to excessive isotope fractionation.

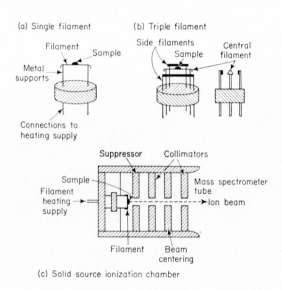

FIG. 4. Solid source filament assemblies.

The maximum efficiency of ionization is produced when multiple filaments are used. With this technique (Palmer and Aitkin, 1953; Tilton, Aldrich and Inghram, 1954; Inghram and Chupka, 1953), the sample is placed on one filament and a second adjacent filament is heated. The heating produces high ionization efficiency of the sample that is being gently evaporated from the cool filament. The use of triple filaments is shown in diagrammatic form in Fig. 4(b).

For some elements such as thorium and gadolinium the production of ThO^+ and GdO^+ is more efficient than that of Th^+, Gd^+ when a single filament is used. In the case of gadolinium, Inghram (1953) has shown that when using a single tantalum filament the calculated value for $n^+/n^0 = 10^{-9}$ at 1250°K.

When gadolinium vapour is in contact with a heated filament at 2500°K the calculated ratio $n^+/n^0 = 3 \times 10^{-5}$, with an increase in ionization of 3×10^4.

In the reaction $GdO \rightleftharpoons Gd + O$ the equilibrium value K equals:

$$K = \frac{[Gd][O]}{[GdO]}$$

From this equation it follows that removal of oxygen by reaction with the filament decreases the oxygen pressure in the equilibrium equation and increases the dissociation.

The use of multiple filament techniques has the added advantage that very small samples may be analysed, and the limit of detection is lowered. When a single filament is heated, the maximum heating occurs in the middle. Kendall (1958) has improved the production of K ions by placing the sample at one end of the heated filament. It vaporizes at the cooler end and is subsequently ionized in the hot central region. It is essential for controlled ion emission that the filament current be stabilized.

While equation (11) states the theoretical relation between the production of positive and neutral ions from a filament, the practical application is limited, as the work function of a particular filament material will vary considerably, as a result of chemical reactions at the sample–filament inter-surface.

Attempts at defining the production of ions from a filament are further complicated by isotope fractionation in the sample as evaporation progresses. The fractionation effects are described by the Rayleigh formula (Cohen, 1951), which states that as evaporation proceeds the sample will be progressively enriched in the heavier isotope. These effects, described by Brewer (1936), Hoff (1938), Reuterswärd (1956), Bentley, Bishop, Davidson and Evans (1959), Shields, Garner, Hedge and Goldich (1963) and Eberhardt, Delwiche and Geiss (1964), show the following.

(i) Measured isotope ratios show a time dependent variation corresponding to a Rayleigh distillation.

(ii) The slope of the Rayleigh curve is dependent upon the chemical form of the sample and the type of filament used.

(iii) The initial stage of evaporation often shows a very large deviation from the Rayleigh curve, which is a critical factor in determining the average isotope ratio. The reason for the initial enrichment is not fully understood and, unless it is reproducible, it will constitute a source of error.

(iv) The observed discriminations are of the same magnitude as the square root of the ratio of the masses.

(v) The highest observed enrichments are found for light elements such as lithium, potassium and rubidium. Eberhardt et al. (1964) have observed enrichment of potassium and lithium of 10% and 45%, respectively; also, ratios obtained from tungsten were consistently lower than those from tantalum filaments.

Until these features are completely understood, it is obviously difficult to compare absolute isotope ratios obtained by different techniques.

(b) *Furnace sources*

The use of a furnace consisting of a chamber in which the sample is heated within the ionization chamber is not widely used in geological dating. An advantage in this method is that a directed narrow beam of ions can be produced from the chamber exit slit. Some of the disadvantages of the furnace method are condensation of the sample within the source region and memory effects.

(c) *Spark sources* (Craig, 1956)

The production of ions by sparking techniques results in the production of ions of a wide energy range. The large erratic energy spread inherent in a spark source requires an analyser system capable of

TABLE 8. Preferred element form for production of ions

Element	Gas source	Solid source	Ion measured
Helium	He	—	He^+
Carbon	CO_2	—	CO_2^+
Oxygen	O_2, CO_2	—	O_2^+
Argon	Ar	—	Ar^+
Potassium	—	K_2SO_4, $KAl(SO_4)_2$	K^+
Calcium	—	CaO, $CaTaO_4$	Ca^+
Rubidium	—	$RbSO_4$	Rb^+
Strontium	—	$SrSO_4$, $Sr(NO_3)_2$ Sr oxalate	Sr^+
Lead	$Pb(CH_3)_4$	PbS, $PbSO_4$, PbO, PbI	Pb^+
Uranium	UF_4	U_3O_8	U^+
Thorium	—	ThO	Th^+

precise energy focusing. A great advantage of this technique is that the sensitivity for different elements does not vary very much, so it is possible to determine the concentration of many elements without the use of individual standards. The mass spectrum is recorded on a photographic plate and is in appearance analogous to that obtained by an optical spectrograph. The method is extremely sensitive for a large number of elements and it would

2*

appear that this method has a great future in studies of the distribution of trace elements in geological material.

It is essential that the construction of an ion source should reduce the possibility of cross contamination of one sample by another. All source materials should be non-magnetic, non-corrosive and of a non-absorptive nature.

The chemical form and the preferred type of ion for measurements of geological materials are given in Table 8.

B. Magnetic Field

After passing from the source region, the beam passes between two slits where it is accelerated by a variable electrical field. The emergent ions all have the same energy Ve where V is the potential difference provided by the electrical field. As the energy is kinetic,

$$Ve = \tfrac{1}{2}\,mv^2 \tag{12}$$

The magnetic field is given by the relation Hev where H is the magnetic field strength, e is the charge of the ion and v is the velocity of the ion. Under this influence, the particles follow a circular path of radius r, with a centrifugal force of mv^2/r. When both the magnetic and centrifugal forces are balanced:

$$Hev = mv^2/r$$

$$\therefore\ e/m = v/Hr \tag{13}$$

By combining equations (12) and (13) and eliminating v,

$$m/e = \frac{H^2r^2}{2V}$$

Under these conditions the magnet acts as a momentum filter and each discrete value of m/e is deflected into a curve of different radius, r. All particles of the same m/e are brought into focus at the exit slit, in spite of having different directions at the entrance slit.

C. Ion Collectors

After passage through the magnetic field the separated ions are collected and converted into an electrical impulse which is then fed into an amplifier.

For large ion currents, a simple metal tube (Faraday cage) is used. The actual design of a Faraday cage varies, but in general it consists of a hollow metal tube which is connected to earth through a high ohmic resistor. As the ion current passes to earth, the potential drop caused in the resistor is a measure of the ion current. The general circuit for this type of collector is given in Fig. 5.

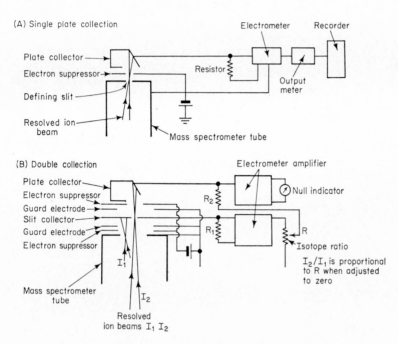

FIG. 5. Single and double ion collection systems.

If the ion currents are small then some means must be provided for "magnifying" the current strength. This is accomplished by use of an electron multiplier. When positive ions moving at speed impinge upon an activated surface (Be–Cu, or Ag–Mg), they produce a number of secondary electrons. The secondary electrons are focused and accelerated on to a second active surface, where they release an even greater number of secondary electrons. Allen (1939) constructed the first electron multiplier, which consisted of 12 stages and, with a potential of 330 V per stage, the primary current was multiplied by a factor of 10^5. The background effects are reduced by insulation and electrostatic shielding. The multiplier is housed within the main vacuum system and the ion current passes directly on to the first multiplier plate (dynode). Two of the essential requirements of an electron multiplier are (a) that the rate of production of secondary to primary electrons must be high and (b) that the activated dynode surfaces should not be impaired or deteriorated by the presence of oxygen. A schematic diagram of an Allen type multiplier is given in Fig. 6.

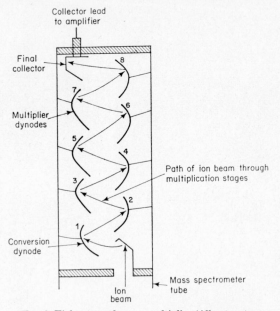

Fɪɢ. 6. Eight-stage electron multiplier (Allen type).

Double collection

In normal isotope studies the electrical and magnetic fields controlling the path of ions allow only ions of one mass to be collected at a time. To obtain isotope ratios it is necessary to measure one, then rapidly bring the other isotope on to the collector. If high precision of measurement is required it is advantageous to measure the two ion currents simultaneously. Straus (1941) was the first to devise such a method, originally proposed by Mattauch and Herzog (1934). Straus measured the ratio of accumulated charge over a period of time. Nier (1947) improved upon this method by making measurements continuous with time (null method).

For double collection measurements the primary defining slit is insulated to function as a second collector. Instabilities in the ion source emission affect ions of the same mass by the same amount and therefore double collection is independent of fluctuations present in the source region. Measurements are made by comparison of galvanometer readings. This method is more accurate for measuring two isotopic abundances that are very similar, but is less accurate for absolute abundance measurements. A general diagram for double collection is given in Fig. 5.

Stevens (1953) has described a variation of the null method of ratio recording suitable for solid source mass spectrometry, which overcomes possible changes in electron multiplier gain, and any isotope fractionation with time. A fine metal wire grid is placed in the ion path and the grid current is compared to the main ion beam detected at the final defining slit. The grid current is proportional to the total emission while that detected at the collector represents successively the currents due to individual isotopes. The grid signal is compared with that from the final collector on a recording potentiometer. A precision of 0·2% can be obtained by this method using surface ionization techniques, but it requires very careful ion beam centring and alignment of the magnet. Although neither the null nor the grid methods are at present used in routine geochronological studies, the former will in the future become essential for measuring very small differences in isotopic abundances.

The final ion current obtained from either the plate or the electron multiplier is fed into either a d.c. amplifier or a vibrating-reed electrometer. In the latter, the current to be measured flows into a capacitor in which one plate vibrates, leading to a change in capacitance. The resultant potential changes caused by the stored charge are amplified by a drift-free a.c. amplifier. The vibrating-reed electrometer has an advantage of high stability over a wide range.

The current from the amplifier passes into a recorder unit. A type in general use is the self-balancing potentiometer utilizing a servo system. The output of the detecting system is compared to a reference voltage. The reference voltage is obtained from a slide wire resistance; the recording pen is attached to the slide wire.

As the current feeding the electromagnet is increased, the pen recorder traces the isotopes (ion current) in increasing mass number. By selecting a small percentage of the mass range, it is possible to continuously move backwards and forwards over a small range. In the case of strontium, the recorded isotopes follow the sequence 88–87–86–84–84–86–87–88. Several methods of measurement of relative peak heights are possible.

D. Vacuum System

The efficiency of a mass spectrometer is very dependent upon the vacuum obtained. An excellent discussion of vacuum practice is given by Martin and Hill (1947) and Turnbull, Barton and Riviere (1962). Mass spectrometers are generally run under continuous flow conditions and the effect of gas evolution from various parts of the spectrometer influences the ultimate vacuum obtained. Very low pressures can be obtained by the use of static flow conditions. Vacuum is obtained in a spectrometer by using mercury or oil diffusion pumps, backed by high-vacuum rotary pumps (Fig. 7).

1. Measurement of pressure

(i) *Tessler coil.* A rough indication of pressure is obtained with a Tessler discharge (induction coil) tube. The coil is placed against a glass part of the vacuum system and the fluorescence produced within the vacuum system indicates the gas pressure. When no fluorescence is visible the vacuum will approximate to < 0.001 mm Hg.

(ii) *McLeod gauge.* This measures absolute vacuum but has the disadvantage that it uses mercury and requires a liquid-air trap so that mercury vapour does not enter the mass spectrometer. The gauge is connected to the vacuum system by tubing and this may lead to a time lag in measurement. The gas pressure is measured by cutting off a part of the gas system into a glass bulb. At the instant of cut-off, the gas is compressed to a known volume in a closed capillary tube. The pressure in the capillary is equal to the difference in height of mercury in the closed and the open capillaries. Permanent gases obey Boyle's Law to 10^{-3}–10^{-1} mm Hg, and this method is of use down to a pressure of about 10^{-5} mm Hg.

(iii) *Pirani gauge.* A Pirani gauge may be used to measure pressure of between 10^{-1}–10^{-4} mm Hg. Its operation depends upon the measurement of the heat conducted by a gas across a space between a heated wire and the walls of the containing vessel. When the pressures reach 10^{-4} mm Hg the conductance of a gas is very small.

(iv) *Thermocouple gauge.* A thermocouple gauge is similar to the Pirani gauge. In this case a flat heater (metal ribbon) carries a constant current. The hotter middle part of the ribbon produces a current in the thermocouple circuit which is registered on a microammeter. The temperature is determined by the heater current and rate of heat transfer from the heater to the glass walls by the surrounding gas. This gauge is used over the range 10^{-1}–10^{-3} mm Hg.

(v) *Penning gauge.* This gauge operates over the range 10^{-3}–10^{-5} mm Hg and utilizes the principle of electrons emitted from a cold surface. The amount of ionization is increased by the use of a longitudinal magnetic field causing electrons to follow a spiral path between two electrodes. The increase in path length by this design increases the total ionization. The small magnet has a field strength of about 400 gauss and the potential between the aluminium or zirconium electrodes is about 2000 V. The ionization discharge results in the gauge acting as a vacuum pump.

(vi) *Ionization gauge.* This gauge is used to measure the lowest pressures in a vacuum system, 10^{-7}–10^{-9} mm Hg. The ultimate vacuum measured depends upon the sensitivity of the meter used. A thin metal anode is heated and outgassed by a spirally wound surrounding filament. When no gas is present, the current produced at the grid is very small. If gas is present, ionization by collision with the heater filament electron stream results in the ions being produced which drift to a negative grid. The current from the grid is a measure of the gas pressure.

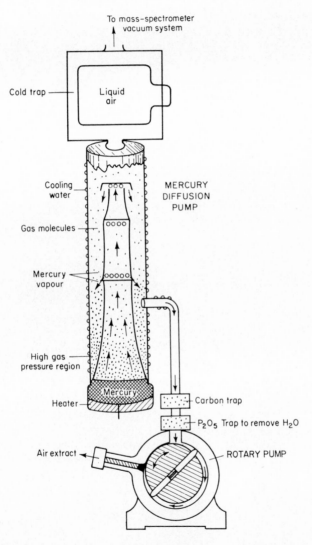

FIG. 7. Diagram of a mercury diffusion pump backed by a mechanical rotary pump. Mercury vapour passes through the holes in the central baffle systems imparting part of their velocity to gas molecules (air), the acquired velocity impels them towards the bottom of the pump where greater pressure conditions allow them to be removed by the mechanical pump.

2. Production of vacuum

Mercury diffusion pumps backed by rotary pumps (Fig. 7) provide the main mechanism for obtaining a vacuum within the mass spectrometer. A diffusion pump consists of a "pool" of mercury that is heated, the vapourized mercury passes up a column and is then condensed by means of a cold trap. Gas molecules present collide with the mercury vapour and the gas released by condensation of the mercury is removed by the backing pump. In some spectrometers mercury is replaced by various oils having very low vapour pressures and which often do not require cold traps.

While mechanical and diffusion pumps are standard equipment on most spectrometers, recently ion pumps have been used. Sputter ion pumps are similar to Penning gauges in which the cathode consists of a titanium plate. Positive ions formed at the cathode discharge are accelerated in an electric field and fall on to the anode forming a fresh surface that will actively attract by chemical reaction gas molecules and atoms. Single cell types have a rather low pumping rate but this can be increased by using a large number of cells in a honeycomb arrangement. An alternative type, the evaporation pump utilizes the evaporation of titanium on to a cold condenser surface—titanium wire is continuously fed against a hot filament where it evaporates and then condenses on a cooled condenser. This type of pump is capable of obtaining pressures of 10^{-7} mm Hg.

Ion pumps have the following advantages. (1) They do not use liquids, are constructed of low vapour-pressure metals and consequently systems utilizing them are free from contaminants such as mercury vapour and organic liquids. (2) Baffles and condensing traps are not required, which would add impedance to the system and reduce pumping speeds. (3) The pumps are more reliable and robust than conventional pumps.

Although glass is sometimes preferred for the construction of mass spectrometers, all-metal machines are common. When used for the analysis of gases in particular, it is necessary to remove any absorbed gas. This is accomplished by removal of the magnet and enclosing the tube within an oven. For this reason all material must be capable of withstanding the baking conditions. It is essential that all parts of a spectrometer are clean, smooth and free from porous areas. Joins in glass parts are made with tight fitting greased joints, while metal parts are sealed by metal or plastic gaskets. Isolation of various parts of the spectrometer and control of gas flow are accomplished by valves.

E. OMEGATRON

The omegatron is a machine suitable for geological work and was originally designed for leak detection. It employs a radio-frequency field. In this system, ions move in a curved path under the control of a radio-frequency electric field and also a magnetic field. Particles move in a spiral path and mass separation occurs. The magnitude of mass separation is increased by increase of ion path length, but this in turn leads to a decrease in resolving power because of gas scattering along the path length. For this reason the ion path is kept as short as possible. Features of the omegatron are its very small size and its strong magnetic and weak radio-frequency fields. It is extremely sensitive, as no slit systems are used and the resolving power varies inversely with ionic mass. The omegatron is useful for mass numbers less than 40, and may prove to be a valuable instrument for the determination of very small amounts of argon and other rare gases in rocks and meteorites, particularly when the sample size is small.

CARBON-14 METHOD

Korff (1940) suggested the possibility that ^{14}C is continuously being generated in the upper atmosphere by cosmic-ray-produced neutrons. In 1946, Libby, using then available data, estimated that it should be possible to detect the radioactivity emitted by ^{14}C in samples taken from the atmosphere or from terrestrial samples that have exchanged with CO_2 from the atmosphere.

Libby and his co-workers (Anderson, Libby, Weinhouse, Reid, Kirshenbaum and Grosse, 1947) then proved by experiment that the radioactivity of ^{14}C could be detected and suggested its use as a method of dating.

Natural modern carbon consists of two stable isotopes ^{12}C and ^{13}C with a natural abundance of 98·892 and 1·108%, respectively, while a third radioactive isotope, ^{14}C, is formed by the reaction of cosmic ray particles (thermal neutrons) with ^{14}N in the upper atmosphere. The abundance of ^{14}C in modern wood is 0·000,000,000,107% which, because of the relatively short half-life, it is not found in old carbonaceous materials such as coal.

Libby (1955) determined a half-life for ^{14}C of 5580±40 years, Olsson, Karlen, Turnbull and Prosser (1962) a value of 5680±40 years, while recently a Dating Conference held at Cambridge in 1962 (Goodwin, 1962) adopted a value of 5730 years. The age of a carbon sample is related to the calculated common disintegration law in the following manner:

$$N = N_0 e^{-\lambda T} \tag{14}$$

where N_0 is the activity of ^{14}C present in the sample at the time at which the sample was removed from the natural carbon cycle. For an organic organism this would date the time at which the organism died. N is the activity of ^{14}C at time T and λ = disintegration constant.

I. Production of ^{14}C in the Atmosphere

Carbon-14 is formed at an altitude of about 1600 m according to the equation

$$^{14}N\ (np) \to {}^{14}C \overset{\beta^-}{\to} {}^{14}N \tag{15}$$

While this reaction is the dominant one, others such as

$$^{16}O\ (n\ ^3He) \to {}^{14}C$$
$$^{17}O\ (n\alpha) \quad \to {}^{14}C$$
$$^{15}N\ (nd) \quad \to {}^{14}C$$
$$^{13}C\ (n\gamma) \quad \to {}^{14}C$$

are of only minor importance.

The conversion of the newly formed ^{14}C atom into a molecule of CO_2 is not fully understood, but of several possibilities it is generally considered that one of the following is the most likely mechanism.

(a) Either a direct reaction with oxygen to form carbon dioxide, or an indirect one to form carbon monoxide which is subsequently oxidized by the ozone layer to form CO_2.

(b) Pandow, Mackay and Wolfgang (1960) consider that a simple oxidation reaction in the atmosphere is not possible, and suggest that the actual conversion of $^{14}C \to {}^{14}CO_2$ is carried out in the soil through the agency of a micro-organism, *B. oligocarbophilus*.

(c) Dorn, Fairhall, Schell and Takashima (1962) have suggested that the

$$CO_2 + {}^{14}CO \to {}^{14}CO_2 + CO$$

reaction is carried out in the atmosphere with the aid of sunlight.

In calculating a radiocarbon age it is assumed that both the rate of production of ^{14}C by reaction with neutrons, and the concentration of N_2 in the atmosphere has been constant in terms of the average half-life of 8033 years. (Average half-life is longer than $t_{\frac{1}{2}}$ by $1/\ln 2$.)

Neutrons bombard the earth's surface at a rate of 2·6 n/sec per cm², amounting to a total neutron flux of $1\cdot3 \times 10^9$ n/sec cm². If all the available neutrons resulted in the formation of ^{14}C this would amount to about 80 metric tons of ^{14}C, which is equal to 360×10^6 curies of activity (1 curie $= 3\cdot7 \times 10^{10}$ disintegrations/sec). From calculations based upon observations, the total amount of ^{14}C found in the atmosphere is approximately 0·95 metric tons, which means that practically all of the ^{14}C is rapidly removed from the atmosphere, probably by downward flowing convection currents.

The intensity of the total cosmic flux varies with altitude and latitudinal position (Fig. 8), reaching a minimum at the geomagnetic equator and a maximum at high altitudes (Simpson, 1948; Lingenfelter, 1963). By determining the present day activity of modern wood over a wide range of both latitude and height, Anderson and Libby (1951) (Table 9) have shown that within narrow experimental limits there are no significant differences in ^{14}C activity and therefore the turnover time for ^{14}C must be short, relative to the mean life of radiocarbon. The production of ^{14}C may not have been constant in the past,

TABLE 9. Specific activity for ^{14}C from different geomagnetic latitudes

Source	Geomagnetic latitude	Absolute specific activity (dpm/g)
White spruce, Yukon	60°N.	14·84±0·30
Norwegian spruce, Sweden	55°N.	15·37±0·54
Elmwood, Chicago	53°N.	14·72±0·54
Fraximus excelsior, Switzerland	49°N.	15·16±0·38
Honeysuckle leaves, Oak Ridge, Tenn.	47°N.	14·60±0·30
Pine twigs, needles (12,000 ft.) Mt. Wheeler, New Mexico	44°N.	15·82±0·47
Briar, N. Agrica	40°N.	14·47±0·44
Oak, Palestine	34°N.	15·19±0·40
Wood, Teheran, Iran	28°N.	15·57±0·31
Fraximus, Mandshurica	26°N.	14·84±0·30
Wood, Panama	20°N.	15·94±0·51
Chlorophora excelsa, Liberia	11°N.	15·08±0·34
Sterculia excelsa, Copacabana 9000 ft. Bolivia	1°N.	15·47±0·50
Iron wood, Marshall Is.	0	14·53±0·60
Wood, Ceylon	2°S.	15·29±0·67
Beechwood, Tierra del Fuego	45°S.	15·37±0·49
Eucalyptus, N.S.W. Australia	45°S.	16·31±0·43
Seal Oil, Antartica	65°S.	15·69±0·30
	Average	15·3 ±0·1

After Anderson and Libby (1951).

and Elasser, Ney and Winckler (1956) regard changes in the earth's magnetic field to be sufficient to reduce the ^{14}C : ^{12}C ratio. The effect of such changes on radiocarbon ages would mean that samples younger than 2000 years will be correct to 250 years.

Using ^{14}C produced by nuclear explosions, Hagemann, Gray, Machta and Turkevich (1959) indicated a mean atmosphere residence time of 5 years. Based upon the residence time of ^{90}Sr in the stratosphere, Feely (1960) suggests a period of 18 months, while the exchange time between the atmosphere and vegetation is very rapid (Broecker, Olson and Bird, 1959). Using fossil fuel (Suess effect, see p. 41), Fergusson (1953, 1955) has shown that the atmospheric mixing rate between the N.–S. hemispheres is less than 2 years. A comparison of variation in the cosmic flux with geomagnetic latitude is given in Fig. 8, and the specific disintegration for ^{14}C is given in Table 9.

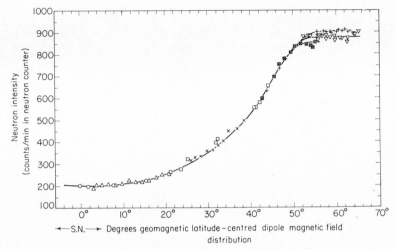

FIG. 8. Latitudinal variation of cosmic-ray neutron intensity (after Simpson, 1948).

It is generally accepted that during the last 8000 years no large changes over prolonged intervals of time have occurred in the average cosmic flux, in the magnitude of the magnetic field in the vicinity of the earth, and in the degree of mixing in ocean water (Libby, 1963). A discussion of geophysical variations has been given by Wood and Libby (1964).

Other errors arise after the carbon has been removed from the biosphere cycle. One of the most common is the contamination of older material by more recent carbon derived from humic acid, root hairs or fungal growth.

II. Methods

It is essential to the method that the carbon to be measured from a sample contains the original atoms present when the sample was removed from the carbon cycle. The covalent bonding of the hydrocarbon molecules are less susceptible to replacement than those of inorganic molecules, but carbonaceous material forms an excellent surface for exchange reactions with ground water. The magnitude of the errors caused by contamination of the sample with recent material is given in Table 10 (Olsson, 1957, p. 58). Anomalous ages obtained by this form of contamination can sometimes be overcome by sampling as many different forms of carbon as possible at one site. The sample is carefully cleaned to remove any adhering material, such as soil or modern organic life, and is then treated with dilute acid to remove any carbonate, after which it is thoroughly washed and dried. Inorganic forms of carbon, such as shells, are treated with dilute acid to liberate carbon dioxide.

TABLE 10. Errors caused by the addition of modern carbon

Age of sample (years)	Error (years) Percentage contamination with modern carbon		
	1%	0·1%	0·01%
5570	80	—	—
11,140	240	20	—
22,280	1200	120	10
44,560	12,300	1800	200

After Olsson (1957).

Chemical and physical procedures employed at this stage depend upon the final counting media required. Libby (1955) has used carbon black, Suess (1954) used acetylene, Burke and Meinschein (1955) used methane and Olsson (1957) used CO_2 gas while, more recently, the incorporation of a carbon compound into a scintillation liquid has been used (Hayes, Anderson and Arnold, 1953). In all cases prior to counting, the sample is left for 2–3 weeks to allow any contaminating radon to decay.

The present day average activity for carbon is $16·1 \pm 0·5$ dpm per gm per C (Anderson and Libby, 1951). The low level of activity emitted by radiocarbon, and the need to have the maximum count rate of sample to counter background, necessitates the use of special low background counters.

III. Counters

An average end-window Geiger counter has a background of about 10 counts/min and a poor counting efficiency of 5–10%. Even with massive shielding to remove some of the cosmic ray components, no simple design of this counter is suitable for radiocarbon dating.

A. SCREEN WALL COUNTER

Libby (1955) and Anderson and Libby (1951) were the first to describe a counter suitable for measuring the amount of emitted radiation. While the end-window Geiger counter has a very poor efficiency, the maximum efficiency can be obtained if the sample is coated on to the internal wall. With the no screen counter the sample is coated on to the inside surface of a cylinder which surrounds a Geiger tube. The normal cathode (wall) of the Geiger tube is replaced by a wire grid which dispenses with the problem of absorption by an end window. The effective geometry is about 50% and the surface area of the sample is large, which in part compensates for the loss of counts by absorption in the carbon sample. Radiocarbon emission has a range of 28 mg/cm² (i.e. a foil weighing 28 mg/cm² will just stop the radiation) so great care must be exercised to ensure very even thickness of the carbon deposit. Libby's counter will accommodate 8 g of carbon over an area of 400 cm² and with an absolute counting efficiency of 5·4% for ^{14}C.

B. GAS COUNTERS

1. Geiger counter

This type of counter consists of a geiger tube in which the sample, in a gaseous form (mixtures of $CO_2 : CS_2$), forms the counting gas. The very high efficiency of counting, which is approximately 100% compensates for the large sample size allowed for in the screen wall counter, but the sensitivity of these systems is rather low. Counters of this type have been described by Crane (1954) and Moscicki (1953).

2. Proportional counter

The use of proportional counters allows a reduction in the number of background counts by the use of pulse height discrimination. Low-background proportional counters using CO_2, C_2H_2 or CH_4 as the counting gas have been described by De Vries and Barendsen (1952), Suess (1954), Fergusson (1955) and Olsson (1957). In the early attempts, poor counting characteristics were associated with gaseous CO_2. However, it has been shown (Fergusson, 1955) that this was not due to the gas, but rather to electronegative impurities in the gas which capture electrons to form negative ions. As the principle of the counter is the utilization of electron collection, this reaction is highly undesirable. Oxygen is one of the most common contaminants; others include water, ammonia and hydrogen chloride. While working under a pressure of greater than 1 atm, the background count rate increases very slowly for the pure gas, but is markedly increased when electronegative impurities are present. Carbon dioxide can be suitably purified by adsorption on to hot calcium metal (de Vries and Barendsen, 1952) followed by gradually increasing the temperature so that compounds of a different dissociation temperature from that of carbon dioxide are preferentially removed. At a temperature of 800°C carbon dioxide is released and stored prior to counting. The optimum counting efficiency for ^{14}C is about 68%.

Gas proportional counters have a counting efficiency of 100% compared to about 6% for solid sample counters. They are also free from the hazard of airborne contaminants and can be used for as little as 0·1 g of carbon.

C. Low-background Counters

The low level of activity emitted by radiocarbon samples, and the need to have the maximum count rate of sample to counter background necessitates the use of special low-background counters. The error quoted on a radiocarbon age is solely an error in counting statistics. While special procedures are adopted for scintillation counters, the background of Geiger and proportional counters may be as high as 50 counts/min. The background of the latter consists of the following components.

Total count rate	=	Activity from C^{14}	+	Activity from radioactive contamination in the sample	+	Cosmic activity detected by counter	+	Radioactive contaminants in equipment materials.

The reduction of the background can be accomplished by shielding. By surrounding the counter with large amounts of iron or lead the hard cosmic ray component is reduced. Kulp (1952) has used an additional layer of distilled mercury to absorb radiation from the iron. The soft component of the cosmic radiation produces neutrons in the iron shield which can be reduced by placing a material of low atomic weight such as paraffin wax mixed with boric acid within the iron shield (de Vries, 1956, 1957).

It is becoming increasingly difficult to obtain iron or steel shielding materials that do not contain appreciable amounts of a radioactive contaminant. In modern iron and steel production, radioactive tracers such as ^{60}Co are added to control mixing during the smelting process; in addition, radioactive waste products produced during nuclear explosions contaminate modern materials. The radioactivity from materials used in the construction of counting chambers is given in Table 11.

An alternative to the use of material absorbers is the use of an anti-coincidence shield which consists of a series of tangentially placed geiger tubes operated in anti-coincidence, positioned within the iron shield, but around the central counting chamber. Radiation that passes through the iron shield is detected by the ring of geiger tubes, the central counting wire is inactivated instantaneously and the spurious radiation is not counted. The combination of iron shielding and an anti-coincidence ring reduces the background of the unshielded counter from say 50 counts/min to less than 1 count/min. The remaining background is mainly derived from radioactive impurities in the wall of the sample counter.

Special counters have been described by Geiss, Gefeller, Houtermans and Oeschger (1958) and by Gefeller and Oeschger (1962) in which the anti-coincidence counters are built into the same tube as the main counter, so that the same gas is used in the whole system. The wall of the counter consists of a polystyrene foil covered on both sides with aluminium. This is then surrounded by a ring of wires forming the anode for the anti-coincident circuit. The thickness of the foil corresponds to 75 keV, which is lower than the maximum energy of ^{14}C. With this system, only about 2% of the ^{14}C activity is lost due to the firing of the anti-coincidence, and this percentage is a constant. Background attributed to α-particles may be removed by electronic discrimination.

TABLE 11. Activities in counter materials

	Disintegrations/min per 100 g	β-Counts/min per 100 cm^2
Lead, modern	350	11
Lead, sixteenth century	35	1
Solder, 30/70 Fluxite	220	7
Solder, 50/50 Wire	700	23
Aluminium 2S	160	6
Aluminium superpure	110	4
Brass rod	60	2
Brass tubing	4, 8, 16	0·1, 0·3, 0·5
Hard solder	50	1·5
Copper, OFHC tubing	1·5	0·05
Polystyrene sheet	0·7	0·02
Polythene sheet	0·8	0·025
Perspex sheet, ICI	0·4	0·01
Plexiglas sheet, Rohm and Haas, UVA	0·7	0·02
Plexiglas sheet, Rohm and Hass, Transparent	1·0	0·03
Lucite rod, Dupont	0·5	0·015
Corning 7750 sealing glass		43
Corning 7052 glass		93
Corning 7740 Pyrex glass		7
Corning 0080 Soda Lime glass		7
Silica		0·3
Steel scale		5
Mica, India Ruby (10 mg/cm^2)		30

After Grumitt *et al.* (1956).

D. SCINTILLATION COUNTERS

There are two types of scintillation counter; one uses a solid scintillation crystal (such as NaI, which is used to detect the radiation from a sample) and the other uses a liquid in which the sample is added in a soluble form (Hayes, Anderson and Arnold, 1956). A "scintillator" has the ability to convert into visible light part of the radiation energy deposited in it by ionizing radiation. The visible light pulses are then detected by a photomultiplier. Solid scintillation crystals such as NaI, CdS, CsI, and LiF have high refractive indices, which causes difficulties in the optical coupling between the sample and the photocathode of the photomultiplier. In addition, the crystal substance deteriorates in the presence of water vapour, so they are normally sealed in a glass or plastic container. In liquid scintillation counting, the solute is responsible for the production of a light pulse and can be used at concentrations as low as 5%. Common solutes consist of aromatic hydrocarbon chain molecules such as terphenyl. The solvents used to dilute the active scintillator are strong light absorbers, so there must be an efficient transfer of energy for the light pulse to reach the solute and this limits the solvents to the alkyl benzene group. It is possible to modify the solvent by the addition of an aliphatic compound such as ethyl alcohol so as not to impair the transference of light. A further improvement can be made by adding a hydrocarbon such as 1,4-di-(2-(5-phenyloxazoyl))benzene (POPOP) to shift the wavelength of the emitted light into the region of maximum response for the photomultipliers (Hayes, 1955, 1956).

The sample is converted into an appropriate soluble form and is added to the scintillation liquid. There are no effects from self-absorption and the counting geometry is essentially 100%. The efficiency of liquid counters is less than that for solid crystals and it is essential to count the pulses resulting from single electron emission from the photocathode of the photomultiplier tube. These pulses are of the same order of magnitude as the thermionic noise pulses of the photomultiplier tube. The effect of these can be removed by refrigeration of the photomultiplier and the use of two photomultipliers directed at the liquid in coincidence. In such a system, the simultaneous appearance of light pulses is registered on the counter, while random thermionic pulses for each tube are rejected.

Liquid scintillation techniques have the great advantage that the total volume of the counting media is small and the sample is in a condensed phase with a density many times greater than gases; it therefore subtends a much smaller angle to the external background. Massive shielding and the use of anti-coincidence shields are not necessary as the cosmic ray particles (mesons) give large pulses in a liquid scintillator and are easily discarded by pulse height discrimination.

The background is mainly determined by the dark current of the phototube, which may be rather high for the energies of ^{14}C dating. While these problems can possibly be overcome by using carefully matched photomultipliers and stable electronics in conjunction with a coincidence circuit, there remains the rather complex synthesis of the sample into a form convenient for incorporation into the liquid scintillator. This process does not normally result in 100% yields and there arises the possibility of isotopic fractionation. Barendsen (1957) has overcome some of these difficulties by using liquid carbon dioxide as a diluent.

IV. The Distribution of ^{14}C in Nature

In time, the radiocarbon formed in the atmosphere (Libby, 1955; Arnold and Anderson, 1957) enters the terrestrial carbon cycle and is distributed throughout the various carbon phases in a homogeneous manner. While it is difficult to calculate the length of time required for complete mixing, observations indicate that in general it is less than the half-life of ^{14}C. Craig (1957) has calculated by the application of steady state equations to the transfer and distribution of ^{14}C between its various exchange reservoirs that the residence period for ^{14}C in the atmosphere is 4–10 years. In comparison, the time required for mixing in surface rock and carbonate sediments will be long. As the cosmic production of ^{14}C is such a favoured reaction it can be approximated that the average incident neutron flux in terms of neutrons/cm per sec is directly related to the production of ^{14}C. The amount of ^{14}C present on the earth must be such that its rate of disintegration equals that of its formation. The exchange reservoir through which ^{14}C is assumed to be mixed in a uniform manner consists of the following.

	g per cm² earth surface
Ocean, carbonate, bicarbonate	7·25
Ocean dissolved organic matter	0·59
Biosphere	0·33
Atmosphere	0·12
	8·3

The average neutron flux incident on the earth's surface has been estimated to be 2·4 cm² sec (Ladenburg, 1952); and, more recently, 2·50 cm² sec by Lingenfelter (1963) therefore the number of disintegrations per minute for the "exchange" reservoir should be:

$$2\cdot4 \times \frac{60}{8\cdot3} = 17\cdot2 \text{ dpm}$$

which agrees closely with the observed rate of 16·1±0·5 dpm. It is now necessary to describe briefly two procedures that can cause significant error in radiocarbon dating.

Isotopic fractionation can occur both in nature and during the chemical procedures employed to extract carbon from organic sources.

For the latter, the amount of fractionation produced in different laboratories has been determined by circulating a carbon bearing standard (oxalic acid) in which the amount of fractionation is determined by mass spectrometric measurement relative to ^{13}C. The amount of laboratory fractionation produces about a 1% variation in the reported activity (Craig, 1961), while fractionation produced in nature is of the order of 10%.

Anderson and Libby (1951) were able to show that the distribution of ^{14}C in modern carbon from a wide range of living matter was essentially constant provided that isotopic fractionation had not occurred in the atmosphere, biosphere or sea. A mass spectrometric study of the ratio $^{13}C : {}^{12}C$ in different materials shows that isotopic fractionation does occur. A fractionation of as much as 6% can occur between wood and carbonate. The amount of fractionation is measured in terms of differences δ between a sample and a standard,

i.e.
$$\delta = \left(\frac{{}^{13}C/{}^{12}C \text{ sample} - {}^{13}C/{}^{12}C \text{ standard}}{{}^{13}C/{}^{12}C \text{ standard}} \right) \times 1000$$

Terrestrial plants show an enrichment in ^{12}C while marine liminetic plants are enriched in ^{13}C with respect to terrestrial plants. The total fractionation observed for plants is not regarded as being produced at one stage but as the result of several fractionation stages within the plant (Craig, 1954). The $^{13}C/{}^{12}C$ ratio for natural carbonate is greater than that for organic material by 2·6% (Craig, 1953). The expected enrichment for ^{14}C in modern shells should be about 5%, but the observed enrichment is zero (Craig, 1954), indicating that the specific activity for ^{14}C has been reduced by 5%, which corresponds to a time of 400 years for bicarbonate in surface waters of the ocean. This illustrates the incomplete exchange of ^{14}C between the atmosphere and the oceans. Carbon-13 enrichment corresponds to a significant error in radiocarbon dating and is far greater than errors caused by isotopic fractionation introduced by the decomposition of organic material, or those caused in the laboratory.

Apart from natural isotopic effects such as fractionation, man has taken a hand in upsetting the natural balance of the ^{14}C reservoir. The explosion of nuclear devices releases neutrons which will alter the ^{14}C content of the atmosphere. Crane (1951) visualized such events as providing a means whereby it would be possible to study the carbon economy of the earth under controlled conditions. Since the end of the nineteenth century man has progressively released to the atmosphere ancient carbon by the combustion of coal and oil. Suess (1955) observed what is now called the "Suess effect", namely that the specific activity of modern wood is slightly lower than expected due to the dilution of carbon dioxide by a contribution from fossil fuels. This effect has been observed to increase near large industrial sites and has been put to advantage to determine the relation between smog and industrial areas.

V. Application of Radiocarbon Dating

Because of its convenient half-life and the common and world-wide occurrence of materials containing carbon, the radiocarbon method has been extensively applied to the fields of archaeology and anthropology. Radiocarbon dating was used to date the Piltdown skull and jaw and showed that while the skull gave an age of 620 ± 100 years the jaw was only 500 ± 100 years old, proving that the find was in fact a hoax (Oakley, 1950; De Vries, 1959; de Vries and Oakley, 1960). More recently the method has been extended to study the rate of sedimentation of ocean sediments, late Pleistocene geology and climate, and also the terrestrial ages of meteorites. The excellent agreement that can be obtained between the age of known historical samples and the radiocarbon age are given in Table 12. However, there are many radio-

TABLE 12. Comparison of ages obtained by radiocarbon dating with material of known age

No.	Sample	Lab.	Expected age	^{14}C age
1	Inca sample	L	444 ± 25	450 ± 150
2	Roman ship	B	1190 ± 3	2030 ± 200
3	Ptolemy	C	2149 ± 150	2300 ± 450
4	Etruscan tomb	B	2600 ± 100	2730 ± 240
5	Tayinot	C	2624 ± 50	2600 ± 150
6	Sesostris	C	3700 ± 400	3792 ± 50
7	Nippur	C	4125 ± 200	4802 ± 210
8	Zoser	C	4650 ± 75	4979 ± 350
9	Hemalca	C	4900 ± 200	4883 ± 200

L, Lamont (Kulp *et al.*, 1951, 1952a).
B, Rome (Ballaria, 1955).
C, Chicago (Arnold and Libby, 1949).

carbon dates that are not in agreement with historical ages and caution is needed in interpreting results. Some of these errors are possibly related to cumulative errors in historical chronology; ages greater than 4000 years are subject to larger errors as this marks the first astronomical fix. The following examples only serve to indicate the wide range of subjects to which radiocarbon dating has been applied.

(a) Emiliani (1955) has shown from a study of deep-sea cores that during the Pleistocene the temperature of superficial waters in the equatorial Atlantic and Caribbean underwent periodic oscillations with an amplitude of about 6°C. The sedimentation rate for the core samples has been determined by radiocarbon dating and the extrapolation of these results suggests an early temperature minimum at about 280,000 years ago corresponding to the first major glaciation.

(b) While radiocarbon dating has proved to be reliable for world-wide dating of some Pleistocene events, such as the 11,400 years old Two Creeks forest bed and its European equivalent, detailed comparisons over wide areas is often complicated by various items such as (i) variable period of time needed for the development of interstadial soils; (ii) differences in the thickness of soils and oxidation-reduction conditions at different localities and (iii) deposition of secondary glacial deposits containing organic remains.

(c) The dating of glacial events, such as the time corresponding to stages of advance and retreat, are extremely complex as so many complex events took place within a short period of 25,000 years. Horberg (1955) has described some of these factors for the Mississippi-Valley region.

(d) The late Pleistocene extinction of some megaform of animal life has been described by Hester (1960–1). The majority of Pleistocene megafauna existed in herds (mammoth, horse, bison, wolf). In N. America most of these forms became extinct about 8000 years ago, but radiocarbon dating indicates that some forms migrated to the south away from the ice, while the Mastodon possibly survived the longest.

(e) Radiocarbon dating has provided quantitative estimates of time required for the formation of varve deposits (de Geer, 1912), and for the comparative chronology of pollen zones.

(f) The age of recent lava flows can be dated by studying the burnt remains (trees) of organic material destroyed by lava flow (Rubin, 1961; Fernald, 1962).

(g) Radiocarbon dates are in good agreement with ages obtained by the ionium method and indicate the constancy of the cosmic ray flux over the past 30,000 years (Kulp, 1953). A comparison of ionium and radiocarbon dates is given in Table 13. As well as old sediments, radiocarbon dating has been used to date recent and modern sediments. Deffeyes and Martin (1962) have studied the ^{14}C activity from dolomites in the Bay of Florida. The dolomite has been regarded by some as having been recently formed, but the samples gave an age of $>35,000$ years, while the sedimentation only started 4000 years ago, showing that the dolomite was derived from old rocks.

(h) The first reliable estimates for the rate of exchange of CO_2 between the atmosphere and the oceans, using radiocarbon as a tracer, by Revelle and

Suess (1957) showed that the residence time of CO_2 in the atmosphere was about 10 years. Recently, Bien, Rakestraw and Suess (1962) have for the first time determined true age differences of water masses by means of the radiocarbon in the dissolved bicarbonate. Ocean water between 40°S. and 40°N. has an age difference of about 400 years. The amount of ^{14}C in surface water has increased significantly since 1958 due to the uptake of bomb produced ^{14}C from the atmosphere.

TABLE 13. Comparison of ages obtained from deep sea core samples by radiocarbon and ionium methods of dating

Core No.	^{14}C Age	Ionium age
P–130–28	1100 ± 400	1500 ± 500
P–130–118	7000 ± 1000	8200 ± 800
P–130–185	$15,000 \pm 1000$	$14,000 \pm 2100$
P–126–24	$16,000 \pm 1000$	$17,500 \pm 2000$

After Kulp and Volchok (1953).

(j) Until recently it has not been possible to date meteorite falls in an accurate manner. The main estimate for the age of a fall has been based upon the geological formation cut by the craters. Apart from the Ider siderite (Henderson and Clark, 1962) which is associated with quite ancient sedimentary rocks, the others are not found in dated stratigraphic horizons. Bauer (1947) was the first to suggest that ^{14}C could be produced in meteorites by cosmic-ray interactions. By radiocarbon dating, Kohman and Goel (1962) have shown that many meteorites are thousands of years old, and some are very much older.

Tree rings have been widely used in historical dating and some of the errors encountered in this work may be caused by the addition of extra rings during one year (Glock and Agerter, 1963). The largest errors occur during the period 3000–3600 years B.P.*

The literature describing ^{14}C dating is continually increasing and new results are reported with descriptions of techniques used in the annual journal *Radiocarbon*, a supplementary issue of the *American Journal of Science*.

The method has an interesting future and, while special methods such as the isotopic enrichment of ^{14}C by thermal diffusion techniques using carbon monoxide, enable ages of 70,000 years (Haring, de Vries and de Vries, 1958) to

* B.P., before present.

be measured, the refinement of the normal procedure will soon make it possible to use very small amounts of carbon, such as those in soils. The use of ^{14}C produced in the explosion of nuclear devices (Plesset and Lather, 1960) opens up a large field for studying rates of mixing, and uptake under controlled conditions.

The accuracy of carbon-14 dates has recently been discussed by Libby (1963) and Wood and Libby (1964).

CHAPTER 4

POTASSIUM–ARGON METHOD

I. Introduction

Potassium consists of three isotopes 39, 40, 41, of which potassium-40 is naturally radioactive and decays to two daughter products, calcium-40, and argon-40, by a branched decay scheme (Fig. 9), in which 89% of the disintegrating nuclei go to calcium-40, and 11% to argon-40. In an average granite the rate of production of argon-40 is about one atom per g every 3 sec. In one million years this gives 10^{13} atoms, which is equivalent to 0·005 ppm by weight, an amount that can be measured with moderate precision by modern analytical techniques.

The potassium–argon method has been reviewed by Birch (1951), Wetherill (1957), Aldrich and Wetherill (1958), Zahringer (1960), Lipson (1958), Folinsbee, Lipson and Reynolds (1956), Wasserburg and Birch (1958) and, more recently, in a general review by Hamilton, Dodson and Snelling (1962). A short book on potassium–argon dating has been written by Gerling (1960), while a detailed description of the potassium–argon method is given by Starik (1961) in the book *"Essentials of Geochronology"*: both these publications are in Russian.

Thomson (1905) and Campbell and Wood (1906) were the first to observe that natural potassium was radioactive, as shown by the emission of

3

β-particles, while Köhlhorster (1930), in addition, observed the emission of γ-radiation. The isotopic composition of potassium was first determined by Nier (1936), and Smythe and Hemmendinger (1937) showed that, of the three naturally occuring isotopes, that at mass 40 was radioactive. In 1937, von Weiszäcker suggested that the abnormally higher abundance of terrestrial argon relative to the other inert gases could be accounted for by the dual decay of ^{40}K to ^{40}Ca and ^{40}Ar. He also predicted that old potassium minerals

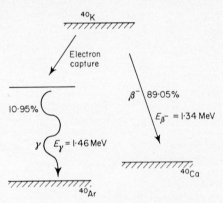

Fig. 9. Decay scheme for ^{40}K.

should contain a large proportion of radiogenic argon, but it was not until 1948 that Aldrich and Nier showed the existence of radiogenic argon in two samples of feldspar and two minerals from an evaporite deposit. Using the ^{40}Ar/^{36}Ar ratio as a guide, they found that the minerals examined contained radiogenic argon, the amount of which increased with geological age. Later, Gerling, Yermolin, Baranovskaya and Titov (1952) published K–Ar ages for micas and feldspars, while Smit and Gentner (1950) dated evaporite minerals.

The dual decay of potassium-40 affords the possibility of two different methods for the calculation of an age. The calcium-40 method has been of little practical use; common calcium is such an abundant element in geological material and the radiogenic ^{40}Ca has the same mass number as that of the most abundant natural calcium isotope, making accurate measurement of the small radiogenic increment difficult, if not impossible. The most suitable calcium poor mineral for this method would be lepidolite from ancient rocks (Holmes, 1932; Brewer, 1938a; Goodman and Evans, 1941; Ahrens, 1951) or a potassium rich mineral such as sylvite found in evaporite deposits. While it is possible that other more common minerals, such as calcium poor biotites, may be suitable, at present the ^{40}Ca decay branch is mainly of value in determining the amount of potassium-40 that decays to calcium-40, for a more accurate determination of the branching ratio.

The potassium–argon half-life of 1·3 billion years is ideal for dating the whole range of geological time, and for young material the only limit is the accuracy with which exceedingly small amounts of the radiogenic argon can be measured. The presence of potassium in geological materials makes the method applicable in principle to almost any geological environment. The rather weak nature of the disintegration energies should not lead to any significant radiation damage of the crystal lattice leading to the continuous loss of argon. In determining a potassium–argon age it is necessary to assume that there has been no loss of argon from the mineral since it was formed. In general the latter requirement is valid provided that the mineral has not been heated at a later date, or there has not been a loss of argon through the process of diffusion. Micas are the most sensitive minerals for potassium–argon dating although, while some feldspars can be used, in the majority of cases argon is lost by diffusion associated with post solidification phase changes.

The isotopic abundances of potassium, argon and calcium are given in Table 14.

TABLE 14. Isotopic abundances of K, Ar and Ca

Element	Isotope	Abundance (atom %)	Reference
Potassium	^{39}K	93·08	Nier (1950b)
	^{40}K	0·0119	
	^{41}K	6·91	
Argon	^{36}Ar	0·337	Nier (1950b)
	^{38}Ar	0·063	
	^{40}Ar	99·600	
Calcium	^{40}Ca	96·97	Bainbridge and Nier (1950)
	^{42}Ca	0·64	
	^{43}Ca	0·145	
	^{44}Ca	2·06	
	^{46}Ca	0·0033	
	^{48}Ca	0·185	

II. Decay Constants

The radioactive decay of potassium-40 produces the stable isotope ^{40}Ca by the emission of a negative β-particle, and ^{40}Ar by K-capture to an excited state followed by the emission of γ-radiation to the ground state.

Although the decay is associated with two partial decay constants λ_β, λ_K, potassium-40 can have only one half-life given by:

$$t_{\frac{1}{2}} = \frac{0.693}{\lambda_\beta + \lambda_K}$$

as, by definition, the half-life is related to the total disappearance of a substance regardless of the mechanism by which it disappears. The ratio of K-capture to β^- emission is called the branching ratio and is expressed as $\lambda_K/\lambda_{\beta^-}$. The validity of the decay scheme (Fig. 9) has been discussed by Morrison (1951); some uncertainty exists in the importance of K-capture directly to the ground state of ^{40}Ar. Tilley and Madansky (1959) have indicated an upper limit of 2×10^{-5} for the ratio β^+ to β^- emission in the decay of ^{40}K which amounts to no more than 2% emission direct to the ground state. For the purpose of geological age calculations, this factor is ignored and the specific γ–activity is taken as equalling the rate of electron capture.

The natural abundance for ^{40}K is known with some certainty; Nier (1950a) obtained a value of $0.0119 \pm 0.001\%$ and Reuterswärd (1951) a value of $0.0118 \pm 0.001\%$. Uncertainties in the abundance of ^{40}K have no significant effect upon a calculated age except in the case of very great ages.

A. Physical Determination

1. Determination of λ_K

The emission of 1.46 MeV γ-radiation is difficult to measure as no suitable γ-emitting standards are available which have a simple decay scheme of an energetic β-particle followed by a single γ, and therefore it is not possible to calibrate the counting equipment over the required energy range of the sample. A determination of the specific γ-activity, corresponding to the electron capture decay, was found by Wetherill (1957) to be $3.30 \pm 0.12 \, \gamma/g$ per sec, corresponding to a partial electron capture decay constant of 5.85×10^{-11} year^{-1}. In this measurement Wetherill made use of accurately standardized ^{60}Co and ^{24}Na sources, thus determining the efficiency of the counting system at 1.17, 1.33 and 1.38 MeV. This allows a measurement of the variation of the efficiency with energy as well as the determination of the absolute efficiency at one energy. The extrapolated efficiency at 1.46 MeV was found to be $0.625 \pm 0.020\%$. A discussion of the various values obtained for the specific γ-activity of ^{40}K is given by Aldrich and Wetherill (1958).

2. Determination of λ_β

As a result of the high β-energy, with a maximum of 1.34 MeV, the problem of self absorption in the counting medium is not serious. The techniques of measurement are similar to those used for the determination of ^{87}Rb (see p. 86), and Aldrich and Wetherill (1958) suggest a specific β-particle of 27·6 β/g per sec, corresponding to a partial β^- decay constant of $4·72 \times 10^{-10}$ year^{-1}. By liquid scintillation counting techniques, Glendenin (1961) obtained a value of $28·2 \pm 0·3$ disintegrations/sec per g. Using the isotopic abundance of ^{40}K as 0·0119 atoms %, and a value of $0·11 \pm 0·02$ for the electron capture this gives a total half-life for ^{40}K of $1·26 \times 10^9$ years.

Alternative forms of the decay constant that are also used are:

$$\text{branching ratio } R = \frac{\lambda_K}{\lambda_\beta} = 0·11;$$

$$\text{half-life} = \frac{0·693}{\lambda_K + \lambda_\beta} = 1·26 \times 10^9 \text{ years.}$$

B. "Geological" Determination

"Geological" determinations of the decay constants by the measurement of total potassium and radiogenic argon from minerals of known age have run into difficulties because of argon loss. In the geological method the branching ratio is determined by comparing results obtained from micas and feldspars to uranium–lead ages obtained from a suitable co-existing mineral. The nature of this mineral assemblage restricts this approach to pegmatites. Mousuf (1952) and Russell (1953) showed that the radiogenic argon obtained from feldspars of known age was consistent with a branching ratio of 0·06. Agreement between uraninite ^{207}Pb/^{206}Pb and muscovite ages from the same pegmatite appeared to indicate that the feldspars had not lost argon to any appreciable extent. Wasserburg and Hayden (1954) repeated this experiment and obtained a significantly greater yield of argon from one of the samples. Later, Shillibeer, Russell, Farquhar and Jones (1954) showed that the low yields in the early experiments were due to the incomplete extraction of argon from the samples. Using young micas associated with minerals having concordant uranium–lead ages Wasserburg, Hayden and Jensen (1956) 'determined a branching ratio of 0·118, while Wetherill, Wasserburg, Aldrich and Tilton (1956) obtained a ratio of 0·126 for argon extracted from a sylvite. Taking into account uncertainties in both λ_β and λ_K the results from geological evidence suggest a branching ratio of $0·118 \pm 0·015$, in which the main source of error is in the value taken for λ_β. From the geological approach, Wetherill

$$\lambda_K = 5·57 \pm 0·15 \times 10^{-11} \text{ year}^{-1} \quad \lambda_\beta = 4·7 \times 0·5 \times 10^{-10} \text{ year}^{-1}$$

$$t_{\frac{1}{2}} = 1·32 \times 10^9$$

et al. (1956) obtain the following. The value λ_K is still a few per cent lower than that obtained from the best counting measurements. This discrepancy has been attributed by Aldrich, Wetherhill, Davis and Tilton (1958) to argon losses from the minerals, in view of the good agreement between the independent counting measurements.

For very young rocks the age is more or less independent of the λ_β-decay constant. In terms of short periods of time the rate of argon production is constant, and the age is equal to the argon content divided by the rate of argon production. For older minerals the sensitivity of the calculated age to errors in λ_β increases, while the sensitivity to λ_K decreases.

The geological age, t, is related to the decay constants by the equation

$$t = \frac{1}{\lambda_K + \lambda_\beta} \ln \left(1 + \frac{\lambda_K + \lambda_\beta}{\lambda_K} \frac{{}^{40}\text{Ar}}{{}^{40}\text{K}} \right)$$

Aldrich and Wetherill (1958) have published curves showing the relative sensitivity of the measured age to changes in the decay constants λ_K and $\lambda_{\beta-}$. From these curves, given in Fig. 10, for an age of 3000 million years, a

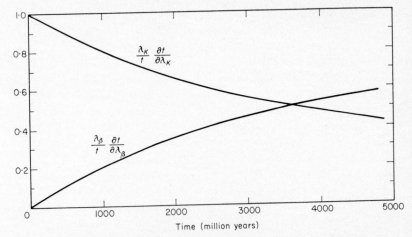

FIG. 10. Relative sensitivity of the measured age t (millions of years) to changes in the decay constants λ_K and λ_β. (From Aldrich and Wetherill, 1958.)

10% error in either λ_K or λ_β will result in about a 5% error in the calculated age. In general, published potassium–argon ages do not take into account any errors in the decay constants and the precision of the analysis is only the sum of the analytical errors.

III. Mass Spectrometry

A major difference between gas and solid source mass spectrometers is that in the former a special system is necessary so that the gaseous sample can be introduced into the source region. The spectrometer may be constructed of glass or metal, although even a glass spectrometer must contain metal parts. Glass is a more suitable material for the construction of apparatus; it can, with due care and experience, be readily and accurately manipulated; it can be rigorously cleaned by chemical means, and gases adsorbed on internal walls can be removed by heating to 300–400°C. The diffusion rates of most gases through Pyrex glass are very small and for argon can be ignored. However, in the analysis for helium, which can easily permeate through Pyrex, Schaeffer (1960) has constructed a special glass system utilizing a high alumina glass which is only 1/10,000 as permeable to helium as Pyrex. Gaskets used to connect together different sections of a glass mass spectrometer must be composed of materials that will not evolve substantial amounts of gas when subjected to conditions of ultra high vacuum; at the same time they should be capable of withstanding bake-out temperatures up to 400°C. Conventional rubber or plastic gaskets are unsuitable and are replaced by metal types, and the gas flow is controlled by special all metal bakeable valves.

Gas that is adsorbed on to glass surfaces may be rapidly removed under vacuum, while gas dissolved in the glass is only slowly removed. Heating the glass walls accelerates the process of desorption and diffusion to the surface, but gas present in deeper parts of the glass is not removed even by prolonged heating. Bake-out procedures are most important in removing gases absorbed in previous analyses, thus removing the "memory effect" of the mass spectrometer; such effects are also caused by the release of imbedded atoms from surfaces bombarded by ions emitted from the sample.

While metal mass spectrometers are of a robust nature and it is possible to accurately position parts of the instrument relative to one another, with the exception of copper, much higher baking temperatures are required to effectively remove dissolved gases. An all-metal ultra high vacuum spectrometer has been described by Signer and Nier (1960). When glass-to-metal seals are required, special graded glass tube is necessary so that the thermal coefficients of expansion can be satisfactorily matched. It is essential that the glass parts of the instrument are completely free from strain, in particular, strain effects set up during the cooling of glass after it has been manipulated. Reynolds (1956) has described an all-glass (Pyrex) gasket-free system to be used in conjunction with ultra high vacuum techniques developed by Alpert (1953). In this instrument, the bore of the bakeable valves limits pumping speeds to less than 0·11 l/sec when all the valves are open but, in routine analysis, pressures below 1×10^{-9} mm Hg are obtained if previously outbaked at 375°C. The absence of gaskets and the use of all-metal valves means that the tube can be outgassed while pumping by enclosing the entire spectrometer in a removable oven.

Electrical conductance on the inner walls of the tube and the removal of surface charge are overcome by coating the walls with a micro-thin layer of electrically conducting tin oxide, formed by heating a film of tin chloride.

For the dating of geological samples the Reynolds (1956) type of instrument is most commonly used. The gas sample is introduced into an ionization chamber through a controlled leak and is ionized by a cross-beam of electrons produced from a heated tungsten or tantalum filament. The stream of electrons is controlled, since the number of atoms ionized is proportional to the filament current. The ion beam emerges from the ionization chamber through a slit and is collimated by passing through a magnetic field of 100 gauss placed parallel to the electron beam. The electron beam is then analysed by passing through a 60° magnetic sector analyser, after which it is registered by a vibrating reed recorder and the output from this fed into a moving chart recorder.

The analysis of a gas sample can be carried out by either dynamic or static operation.

The sensitivity under conditions of dynamic operation is given by the ratio of ions per second reaching the collector to the number of atoms per second passing through the tube (Reynolds, 1956). Under these conditions the sample is continually passed through the instrument so as to maintain a gas pressure within, although there will inevitably be a slight drop in pressure due to the consumption of part of the sample. If the major part of the sample is used up, this leads to isotopic fractionation with a gradual increase in the concentration of the lighter isotope. The fractionation change varies in a linear manner with time and can be corrected for in the final measurement of the isotopic ratios. However, in a dynamic analysis, even if the total sample passes through the ionization chamber, only a small proportion of the total number of molecules is ionized, while the major part of the sample passes out through the pumping system. Within limits, the sensitivity can be increased by using a higher ionization current.

The sensitivity under conditions of static operation is given by the number of ions per second reaching the collector to the number of atoms of that particular isotope in the tube. A small fraction of the total sample is admitted through a controlled leak into the spectrometer tube, which is then isolated from the pumping system. The analysis is then carried out on the isolated portion. The highest sensitivity is obtained under static conditions, which are essential for the analysis of very small amounts of gas. For the latter, ultra high vacuum techniques are required so that a rise in pressure caused by general outgassing is prevented or depressed such that the sample emission is not lost in a tube background. Reynolds (1956) obtained a pressure of 5×10^{-11} mm Hg within 48 h by alternating a period of pumping while baking the tube at 375°C followed by periods at room temperature and pumping with the ionization gauges on. Over a period of 3 h the tube pressure gradually increased to 5×10^{-10} mm Hg as a result of outgassing caused by the ion beam. In the case of xenon, Reynolds has reported a sensitivity of 5×10^5 molecules of xenon and even this could be improved upon by reducing the noise level in the multiplier detector.

Isotopic fractionation in a static analysis is caused by the more rapid passage of the lighter isotope into the tube relative to the heavier isotope. The fractionation is according to Graham's Law, and can be corrected for by multiplying the final measured ion currents by the square root of the molecular weights. When the bulk of the sample passes into the tube this correction is not necessary as there will be some back diffusion of the lighter isotope or, if sufficient time is allowed to pass, a state of equilibrium will be reached.

IV. Separation and Determination of Argon

A. EXTRACTION FROM ROCKS AND MINERALS

Minerals are separated from rocks by normal procedures but, unlike the rubidium–strontium method, it is not desirable to crush a mineral to a very fine powder. In particular, for biotites, the reduction of the mineral to a fine powder grain size leads to the trapping of substantial amounts of air, resulting in large amounts of atmospheric argon.

Argon is removed from the sample either by direct heating or by using a flux. Various fluxes such as Na, $Na_2B_4O_7$, NaOH, Na_2CO_3 and Ca have been used. Shillibeer and Russell (1954) have shown that when sodium is used as a flux the release of argon is dependent upon the grain size of the sample.

In the direct fusion method, the sample is heated in a molybdenum crucible by means of a radio-frequency generator or by direct electrical heating, with either an air or a water cooled furnace as shown in Fig. 11. The evolved gases are then purified to remove compounds that would cause isobaric interference. The extracted gases are first passed over copper oxide in a furnace. Hydrogen and carbon monoxide are oxidized to water and carbon dioxide which are subsequently removed by a cold trap. Following this, nitrogen is removed by allowing the gas to remain in contact with heated titanium or calcium; hydrocarbons are finally removed by thermal decomposition by contact with a heated tungsten filament.

Carr and Kulp (1955, 1957) have investigated the behaviour of very small amounts of argon in high vacuum systems as encountered in potassium–argon dating. In these experiments they used radioactive argon-37, with a half-life of 35 days, and showed that (a) in the furnace part of the apparatus argon is not lost to any appreciable extent by absorption into the glass walls or

3*

on stopcock grease; (b) argon is not lost to the walls of the furnace during radio-frequency heating or to the mercury diffusion pumps; (c) during the fusion procedure argon is completely removed from the sample. These experiments show that quantities of argon as small as 10^{-9} ml can be

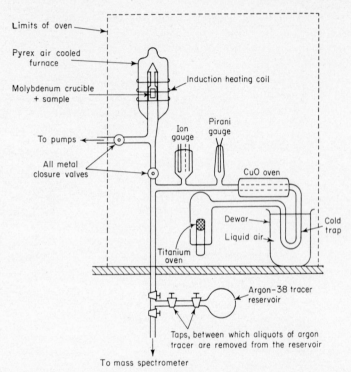

Fig. 11. An argon extraction apparatus. Samples are introduced through the top of the furnace which is then sealed.

circulated in a vacuum system containing a rock melt produced by radio-frequency heating and can be purified by chemical means with a loss of less than 2% of the total argon.

B. DETERMINATION OF ARGON CONTENT OF SAMPLE

1. The total volume method

In the total volume method the gases evolved from a heated sample are chemically cleaned and the remaining chemically inert gases are measured in a calibrated McLeod pressure gauge. The gas is then recycled through the cleaning stages and remeasured to determine whether or not there has been any reduction in the final volume. It can then be assumed that the gas has been thoroughly cleaned and consists solely of argon. This method has yielded acceptable results, but is subject to the following errors.

(a) A systematic error due to faulty calibration in measuring the final volume of gas.

(b) The loss of argon during the chemical cleaning of the gas.

(c) The incomplete removal of contaminating gases during the cleaning stages.

Should these errors be reduced to an insignificant level there still remains an error caused by the presence of atmospheric argon which, in general, amounts to 10–20% of the total argon. A refinement of the total volume method can be made by determining the isotopic composition of the separated argon and use of the $^{36}Ar/^{40}Ar$ ratio to correct for atmospheric contamination.

2. Isotope dilution

Most laboratories determine radiogenic argon by this method as it is less susceptible to errors apart from those of inaccurate tracer calibration or failure to extract all the argon. Some systems require the use of a complex system of glassware, but simpler techniques can be used with equal success. Instead of having to join to the sample system a tube containing the argon tracer, accurately known volumes of tracer can continuously be removed from a reservoir by trapping a reproducible volume of argon tracer between two mercury stopcocks. By isotope dilution techniques the quantities of argon measured range from 10^{-2} ml (s.t.p.) down to 10^{-7} ml or even lower. In comparison the total volume method has a lower limit of 10^{-4} ml (s.t.p.), unless special micro-techniques are adopted. In analysis by isotope dilution the tracer is added to the sample system prior to heating. The sequence of events for the analysis of argon in geological material is as follows.

(1) Load weighed amount of sample into the crucible.

(2) Remove occluded gas by bake-out at 200°C for 12 h. The evolved gases are pumped away.

(3) Test apparatus for leaks.

(4) Admit aliquot of tracer to sample system.

(5) Heat sample when evolved gases are chemically cleaned in CuO, Ti furnaces. Hydrocarbons destroyed by heated tungsten filament. Removal of H_2O, CO_2 in cold trap.

(6) Determine background of mass spectrometer (e.g. mass 28–CO,N_2; 36–HCl; 44–CO_2.

(7) Admit sample to spectrometer through a controlled leak valve.

(8) Measure $^{38}Ar/^{40}Ar$ ratio.

(9) Close valve, pump sample away, measure 36 background (see p. 60).

(10) Re-admit sample, measure ^{36}Ar/^{38}Ar ratio.

(11) Close valve pump sample away, measure 36-background.

(12) Measure peak height and calculate required isotope ratios.

The number of moles of radiogenic argon is determined from the isotopic analysis of the argon according to standard procedures.

The choice of tracer varies in different laboratories, argon-38 is generally preferred although argon-36 or even atmospheric argon (Amirkhanoff Brandt and Bartnitsky, 1961) can be used. Argon-38 is prepared by neutron irradiation of potassium chloride or by gaseous diffusion enrichment (Clusius, Schumacher, Hurzeler and Hosteltler, 1956). Recently, Naughton (1963) has suggested the use of ^{39}Ar, which is prepared by the reaction ^{39}K(np)^{39}Ar in a potassium-bearing glass. The tracer is then added to the sample in the solid form.

The tracer calibration procedure consists of mixing accurately known amounts of tracer and air and determining the isotopic composition of the mixture by mass spectrometry. In a tracer preparation system described by Wasserburg and Hayden (1955), a series of tubes of known volume are simultaneously filled with argon-38 at the same pressure and temperature. After the system has reached equilibrium, the glass tubes are removed and two or three of them are mixed with accurately known volumes of air and the volume of tracer contained in the tubes is calculated according to the equation

$$^{38}V_s = V_a \frac{^{38}\text{Ar}}{^{40}\text{Ar}} \cdot {}^{40}\text{a}$$

where $^{38}V_s$ is the volume of ^{38}Ar tracer, V_a is the volume of atmospheric argon, ^{38}Ar/^{40}Ar is the ratio of isotopes in the calibration mixture and ^{40}a is the fraction of ^{40}Ar in the atmosphere (i.e. 0·996%).

From the determination of the total number of moles of radiogenic argon and the potassium content of the mineral, the equation used in calculating the age of the sample is

$$t = \frac{1}{\lambda} \ln \left[1 + \frac{^{40}\text{Ar}}{^{40}\text{K}} \left(\frac{1+R}{R} \right) \right]$$

where t is the age in millions of years, λ is the total decay constant of ^{40}K $= \lambda_\beta/\lambda_K$, ^{40}Ar is the number of moles of radiogenic argon from the sample, ^{40}K is the number of moles of ^{40}K in the sample (atom fraction ^{40}K in potassium = 0·000119) and R is the branching ratio, λ_K/λ_β (0·584 × 10^{-10} year^{-1}/ 4·72 × 10^{-10} year^{-1} = 0·1237).

V. Determination of Potassium

Although many rocks and minerals contain major amounts of potassium, still it requires much care to obtain results of both high precision and accuracy. To an approximation, an error of 1% in potassium content determination results in an error of 1% in the final potassium–argon age.

A. CLASSICAL METHOD

Until quite recently the majority of potassium analyses were carried out using the wet chemical J. Laurence Smith procedure (Smith, 1871; Washington, 1930; Groves, 1951). The total mixed chlorides are weighed and potassium is separated from sodium by precipitation with perchloric acid, chloroplatinate or sodium tetraphenylboron (Belcher and Wilson, 1955). This quantitative separation is very time-consuming, requires much skill and the results are generally systematically low through occlusion of potassium on precipitates and loss when volatilizing ammonium salts. These drawbacks are absent in flame photometric methods.

B. FLAME PHOTOMETRY

The flame photometer is a most powerful analytical tool in the determination of alkalis. The precision and accuracy is better than 1–2%, and with careful work this can be reduced to 0·5%. The recognized importance of this method of analysis has led to the publication of three books dealing exclusively with flame photometry (Marti and Muñoz, 1957; Herrmann and Alkemade, 1960; Dean, 1960).

1. Principles

When a solution containing potassium is dispersed in a hot flame the atoms and molecules absorb energy from the flame in passing from the ground state to the higher excited state. Conversely, when the atoms or molecules return to the ground state energy is liberated in the form of a characteristic radiation. This change is expressed in the equation

$$E_x - E_0 = h\nu$$

where E_x is the energy of excited state, E_0 is the energy of original ground state, h is Planck's constant and ν is the frequency of the radiation. The characteristic radiation emitted by the flame passes through a filter or prism on to a photocell, where the radiation is converted into an electrical impulse, amplified and then passed into a recording instrument. Excellent results are obtained with the simply designed EEL flame photometer, the principles of which are given in Fig. 12.

For the determination of potassium, a relatively cool flame is to be preferred because at low temperatures the alkali metals emit abundant radiation while other elements are not excited to anything like the same amount. Common fuel mixtures and their respective temperatures are given in Table 15.

The radiation detected by the photocell consists of the following:

(i) *Simple lines.* Well defined lines which may subdivide depending upon the resolution of the instrument. Apart from flame lines emitted from the neutral atom, the alkali metals and earths show emission from single ionized atoms. The spectra of neutral atoms and ions are quite different.

(ii) *Band spectra.* Generally represented by oxide bands of the alkali metals and earths. At high temperatures there is partial dissociation of gaseous molecules into constituent atoms.

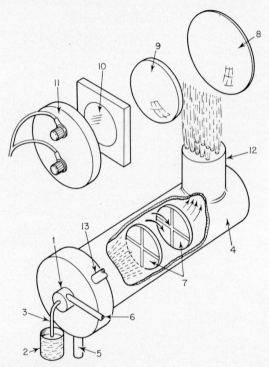

FIG. 12. EEL flame photometer. Air is introduced around the atomizer nozzle through the inlet (6) at a controllable pressure, indicated by a gauge on the front of the instrument. Gas enters the mixing chamber through a tube (13) and is likewise controllable. The airflow around the atomizer is applied to draw the sample (2) up the capillary (3) and through the atomizer, which injects it in a fine spray into the mixing chamber (4), larger droplets falling from the air-stream to pass through the drain tube (5). The baffles (7) produce an even mixture of gas, air and sample, which passes to the burner (12) and is ignited, to burn with a broad flat flame within a well ventilated chimney. Light from the flame is collected by the reflector (8) and focused by the lens (9) on to the photocell (11), the critical wavelenth being isolated by the interchangeable filter (10). The photocell output is then applied through a "sensitivity" control to deflect a built-in taut-suspension reflecting galvanometer in proportion to the intensity of the emitted light.

(Reprinted by permission from Evans Electroselenium Ltd.)

(iii) *Continuous radiation.* Represented by a general background effect as a result of internal electron transitions. The intensity varies with wavelength and for some elements, such as zinc and molybdenum, it reaches a maximum.

TABLE 15. Temperature of fuel mixtures used in flame photometry

| | Temperature (°C) | |
Fuel	Air	Oxygen
Acetylene	2100–2400	3100–3200
Hydrogen	2000–2100	2500–2700
Coal gas	1700–1800	2700–2800

The potassium radiation is measured using the potassium doublet at 766–769 mμ which is free from line interference except for rubidium at 780 mμ and a lanthanum oxide band at 741–791 mμ. The close geochemical and chemical coherence of potassium and rubidium means that the measured potassium includes some radiation from rubidium. In practice this interference is quite negligible and would have to be considered seriously only when analysing a rubidium-rich sample containing minor amounts of potassium, a situation that does not arise in most geochemical studies.

TABLE 16. Ionization potentials and excitation energies for potassium and rubidium

Element	Wavelength (mμ)	Ionization potential (eV)	Excitation energy (eV)
K		4·32	
	766–769		1·60
	404·4		3·05
	694		3·40
Rb		4·16	
	780		1·58
	795		1·55
	420·2		2·94
	421·6		2·93

The similar properties of both potassium and rubidium radiation can be seen in Table 16.

2. Chemical analyses

(Vincent, 1960; Osborn and Johns, 1951; Cooper, 1963; Woldring, 1953; Horstman, 1956; Jenkins, 1954; Margoshes and Vallee, 1956; Abbey and Maxwell, 1960; Easton and Lovering, 1963.)

About 20–200 mg of finely powdered sample are dissolved in a mixture of sulphuric and hydrofluoric acids and made up to a volume of about 100 ml.

(a) *Direct method*

The final volume is made up to 250 ml and separate aliquots are measured directly on the flame photometer. This method relies upon the relative dilution of all possible interfering cations to a level at which they cause no interference to the emission of potassium.

(b) *Chemical separation method*

(i) To the 100 ml of the rock solution a little ammonium carbonate is added to precipitate Fe, Al, Ti, Mg, Ca cations. The solution is then made up to a total volume of 250 ml and aliquots of the clear supernatant are measured on the photometer. In samples containing large amounts of iron and aluminium (basic rocks, biotites) it is necessary to carry out a re-precipitation step to remove occluded or absorbed potassium. This is of particular importance when analysing for small amounts of potassium in basic rocks when larger sample weights are needed for analysis.

(ii) Potassium may be separated from most other elements by ion-exchange methods, but great care must be exercised to ensure complete chemical yields.

(c) *Internal standard techniques*

Lithium is commonly used as an internal standard for the analyses of the alkali metals. The concentration of the standard should be of the same order as the potassium being determined. The sample should not contain any appreciable amounts of lithium and the internal standard must be in a highly pure chemical form. For geological problems this method does not appear to offer any improvements and has the disadvantage that it adds unknown parameters that can affect the final result.

(d) *Standard addition method* (Beukelman and Lord, 1960; Fornaseri and Grandi, 1960)

(i) *Single addition.* A known amount of standard potassium is added to one of two identical aliquots of a sample solution. The concentration of both solutions is determined by reference to a standard linear plot of concentration against emission reading. The difference between the emission of the two samples should give the amount of potassium added. If quenching or enhancement has affected the added potassium, then this will affect the sample to the same amount.

If E_1 = emission intensity of unknown sample (A).

E_2 = emission intensity of unknown sample plus an increment of standard potassium solution (B).

K = background (sample blank + background radiation).

S = amount of standard solution added.

X = concentration of unknown.

$$(E_1 - K)_A = CX \text{ found}$$
$$(E_2 - K)_B = C(X + S) \text{ found}$$
$$E_2 - E_1 = CS \text{ found}$$

where
$$C = \frac{\text{emission of } S \text{ added}}{\text{emission of } S \text{ found}}$$

(ii) *Multiple addition* (Willard, Merritt and Dean, 1958). In this technique at least two increments of standard potassium solution are added to two identical sample solutions. The potassium content of these solutions is then compared with

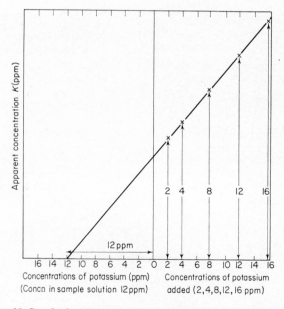

FIG. 13. Standard addition plot for the determination of potassium.

a third pure sample solution. The dilution factor for all three solutions must be identical and the net emission (E_2) is plotted against the concentration of the sample solutions (X). The amount of potassium in the sample is determined by the intercept in the x-axis. Any enhancement or quenching affecting the sample affects the standard in a similar manner and can be corrected.

This technique can be applied over a linear range of concentration, provided that the solutions are dilute so that self absorption effects are not encountered.

An example of this technique is given in Fig. 13 (Evans, 1963).

Sample 610. *Biotite*   (1)   (2)

Weight   0·1067 g   0·09755 g

Dissolution $HF/HClO_4$

Volume   1 litre demineralized   1 litre containing
                          water   15 ml 0·5 N HCl

Method:

    10 ml sample solution placed in beakers A, B, C
    10 ml H_2O added to A
    10 ml $6\gamma K^+$ added to B   ($1\gamma = 1$ ppm $= 1\mu g/g$)
    10 ml $12\gamma K^+$ added to C

Measure the potassium emission of solutions A, B, C against a potassium nitrate standard solution made up in intervals of 1γ.

Results:

	Measured potassium	
	(1)	(2)
A =	3·494	3·170
B =	6·461	6·088
C =	9·470	9·000

Extrapolate to zero concentration (see Fig. 13)

Concentration of K^+ in 1 litre solution

                    $7·118\gamma K^+$ $6·512\,\gamma K^+$
                    0·005   0·005
                    7·113   6·507

$$\gamma K^+ = \frac{\text{concn} \times \text{vol}}{\text{wt}} \times 100$$

$$= \frac{7·113 \times 1000}{0·1067} = \frac{6·507 \times 1000}{0·09755}$$

$$= 6·67\% K^+ \qquad = 6·68\% K^+$$

C. NEUTRON ACTIVATION ANALYSIS

(Smales, 1949; Jenkins and Smales, 1956; Leddicotte, 1956, 1961a, 1962; Salmon, 1957; Smales and Webster, 1957; Lyon, 1964; Mapper, 1960; Winchester, 1961.)

When a rock or mineral is irradiated by neutrons produced in a nuclear reactor, most of the elements become radioactive. A reaction of the neutron–gamma (n–γ) type occurs when an element absorbs a neutron and in the process a γ-ray is liberated. No transmutations occur in a n–γ reaction and consequently the radioactive nuclide has the same chemical characteristics as the original parent element. In general, the intensity of the induced radioactivity is proportional to the weight of the element sought, and is independent of the chemical combination of the element. By irradiating standards and samples simultaneously and assuming that the neutron flux is constant through all the samples, then the induced radioactivity formed in both is directly related to the amount of element present. After the irradiation, the samples are left for an appropriate length of time to allow the short lived activities to decay. They are then chemically processed to separate in a pure form the element to be determined. The "activated" element decays with a characteristic radiation and the half-life is measured by some form of radiation counting. The comparative sensitivity in terms of micrograms of potassium per millilitre for neutron activation, optical spectroscopy data and their relative merits are given in Table 17.

TABLE 17. The determination of potassium

Method	Neutron activation	Flame photometry	Optical spectrograph Copper spark	Graphite d.c. arc
itivity g/ml)	(neutron flux 5×10^{11} 0·080) (neutron flux 1×10^{13} 0·004)	0·01	0·1	
aratus required	Neutron source (reactor) Radiochemical laboratory Counting equipment Trained radiochemist	Simple flame photometer Ordinary laboratory Recording equipment an advantage Unskilled technician	Optical spectrograph Special laboratory Instrument for measuring density of optical lines Skilled technician	
required for icate analysis uding dis- ion if sample aration)	1 h	15 min	1 h	
sion	±0·5%	±0·5%	±1%	±3%
ul concentra- range	<1%	>1%	>1·0%	

Depending upon the potassium content of the sample, between 10–100 mg of powder are weighed into a polythene tube after which the tube is sealed. Between 10–20 samples, together with standards, are then wrapped in aluminium foil and placed in an aluminium can, closed by a screw cap. The can is then irradiated for about 12 h, but this will vary a little depending upon the potassium content of the sample and available neutron flux. After the irradiation the sample is left for a few hours to allow short lived intense radiation of ^{31}Si ($t_{\frac{1}{2}}$ 2·65 h) and ^{28}Al ($t_{\frac{1}{2}}$ 2·27 min) to decay away; for most geological material the bulk of the remaining activity is from ^{42}K($t_{\frac{1}{2}}$ 12·5 h) and ^{24}Na ($t_{\frac{1}{2}}$ 15 h). The relative abundances of the isotopes in natural potassium are: ^{39}K, 93·0%; ^{40}K, 0·0119%; ^{41}K, 6·9%. The decay scheme for neutron irradiated potassium is as follows.

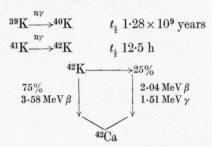

$$^{39}K \xrightarrow{n\gamma} {}^{40}K \qquad t_{\frac{1}{2}}\ 1·28 \times 10^9 \text{ years}$$

$$^{41}K \xrightarrow{n\gamma} {}^{42}K \qquad t_{\frac{1}{2}}\ 12·5 \text{ h}$$

Subsequent treatment of the samples will depend upon the type of equipment available, and to some degree on the bulk chemical composition of the sample. One of the following alternatives is commonly used.

(i) *No chemical separation*. If the main radioactive contaminant is ^{24}Na, this can be removed by covering the sample with aluminium absorbers, filtering out the sodium activity and counting the more penetrating 3·58 β radiation from ^{42}K. In other cases, contaminating activities may be removed by discrimination if γ-spectrometric methods are used.

(ii) *With chemical separation*. It is obvious that if the total potassium from samples weighing 10–100 mg has to be quantitatively separated prior to counting the induced activity, the method of neutron activation has no great advantage over conventional wet chemical methods. However, unlike these methods, after the irradiation the potassium content of the sample is labelled by its radioactivity. In neutron activation procedures, after the irradiation, the sample is brought into solution and a known weight of *inactive* potassium is added so that in the chemical treatment the very small amount of labelled potassium in the sample is carried on the inactive potassium. Once carrier and sample have been completely mixed in solution, subsequent chemistry need not be quantitative as long as some potassium is finally separated in pure form. By weighing the final potassium salt the amount of potassium lost can be calculated. From the chemical yields, the total radioactivity, due to potassium, can then be corrected for a 100% yield.

The same process is often used in chemical separations where a radioactive tracer is added to an inactive sample and the final chemical yields determined from the amount of activity lost during the chemical processing. The great advantage of both methods is that the chemistry need not be quantitative and therefore specific separation techniques may be used.

Chemical separations used in potassium analyses are quite simple and the analysis of 10 samples takes about 2 h. The sample plus carrier is dissolved in a mixture of hydrofluoric and perchloric acids and potassium perchlorate separated. This is then dissolved and, after the addition of iron carrier the hydroxides are precipitated, leaving the potassium in the supernatant.

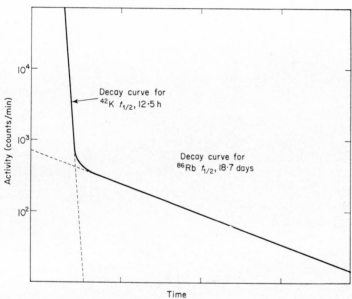

Fig. 14. Composite decay curve for potassium separated from a rock that contained rubidium.

Following this, potassium is separated by cobaltnitrite and chloroplatinate precipitation, the latter being used for the final precipitation. Radioactive purity is checked by following the decay curve, or by γ-spectrometry. The only common contaminant that is not generally separated from potassium will be rubidium. As its half-life is long (19 days) the effects of this contaminating activity can be removed by graphical analysis (Fig. 14).

The radioactive decay of potassium liberates four β-particles for every single γ-decay and, by β-counting, the sensitivity of the method is increased, provided that the β-activity is only due to potassium. With conventional apparatus there is a limitation on the amount of discrimination that can be applied to β-counting. A γ-ray spectrometer can discriminate between different γ energies, but at the expense of total counting rates.

The concentration of potassium in the sample is calculated from the following:

$$\frac{\text{Wt. of potassium in sample}}{\text{Wt. of potassium in standard}} = \frac{\text{Activity of potassium in sample}}{\text{Activity of potassium in standard}}$$

With suitable radiochemical training it is possible to deal with initial high levels of radioactivity provided that certain basic precautions are taken, in particular those concerning health physics. While the activity from samples immediately after irradiation may be very high, the final activity from the separated element may amount to only a few hundred disintegrations per minute.

The counting apparatus required for potassium analysis is relatively inexpensive and simple. A large range of geochemical problems can be tackled with a simple end-window Geiger-Müller counter in conjunction with an e.h.t. supply, probe unit and scaler. It is now becoming common practice to carry out the time consuming procedure of counting many samples by automatic procedures.

The accuracy of this method of analysis (closeness of the mean value to the truth) is difficult to state with certainty as it can only be found by the repeated analysis of a sample in which the trace element in question is known, a situation not generally encountered in geochemistry. The accuracy of the method at trace levels has been quoted as between $\pm 2\%$ and $\pm 10\%$. The precision (reproducibility of repeated individual analyses) of analysis is more easily accounted for and a coefficient of variation of $\pm 0.5\%$ can be obtained. The precision is largely controlled by the statistics of counting.

The method of neutron activation analysis has the following advantages. (1) High sensitivity: the ultimate in sensitivity will depend upon the sample weight, the neutron flux and, within limits, the length of irradiation. (2) The purity of the separated radionuclide can be accurately determined by decay and energy measurements. (3) In theory the method is free from contamination, although it can occur through the addition of the element being determined to the sample before irradiation, e.g. during weighing, through the addition of a carrier or an element that is precipitated with the carrier in significant amounts to affect the chemical yield, or through the incorporation of other radioactive nuclides into the final counting material. Also cross-contamination of a low activity sample with one of a high activity can occur. A general scheme for neutron activation analysis is given in Fig. 15.

Potassium-40 is naturally radioactive with a half-life of $1 \cdot 28 \times 10^9$ years.

The potassium content of geological material can be determined by counting the natural radioactivity, but requires large sample weights and special equipment (Hurley, 1956; Murray and Adams, 1958; Sano and Nakai, 1961).

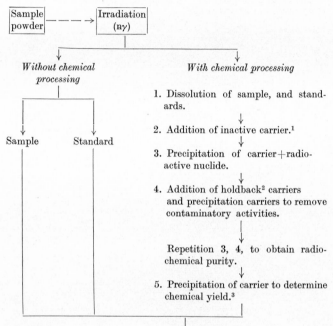

1. Dissolution of sample, and standards.

2. Addition of inactive carrier.[1]

3. Precipitation of carrier+radioactive nuclide.

4. Addition of holdback[2] carriers and precipitation carriers to remove contaminatory activities.

Repetition 3, 4, to obtain radiochemical purity.

5. Precipitation of carrier to determine chemical yield.[3]

6. Identification of radioactive nuclide and resolution from any active contaminants by γ-spectrometry or β-particle counting.

Calculation of result

FIG. 15. General scheme for neutron radioactivation analysis.

[1] Carrier: a known amount of the element being sought is added in a non-radioactive state to the solution of the irradiated sample as a specific carrier of the element sought.

[2] Holdback carrier ⎫ The addition of inactive elements to precipitate contaminating
   Precipitation carrier ⎭ elements, or to retain such elements in solution while the specific carrier is separated. Many elements present in very small amounts at high dilution cannot be chemically processed by normal methods. The addition of macro amounts of various elements allows them to be removed by normal chemical reactions.

[3] The amount of carrier remaining after semi-quantitative chemical separations is directly proportional to the chemical yield of the element being determined.

D. OTHER METHODS

(a) *Isotope dilution.* While this method has the greatest potential for both accuracy and precision, potassium contamination during the chemical processing and in the mass spectrometer ion source is often a serious source of error.

(b) The optical spectrograph and atomic absorption spectrometer (Ahrens and Taylor, 1963; Robinson, 1962) require considerable skill and experience to obtain reliable potassium results, although both instruments are of importance in preliminary reconnaissance work.

(c) Potassium in meteorites has been determined by a distillation procedure (Edwards and Urey, 1955).

VI. Neutron Activation Method for Measuring Potassium and Argon

The determination of the potassium–argon age of geological material by neutron activation was first used by Curran, Dixon and Wilson (1952), Molyk, Drever and Curran (1955), Stoenner and Zähringer (1958a) and, more recently, by Wänke and König (1959) and Wänke (1960). It should be possible to obtain an accuracy of 5–10% by this method, which is sufficient for the dating of large areas where it is at first necessary to determine general age patterns.

This type of analysis is inexpensive compared to that of mass spectrometry, and most laboratories are within reach of nuclear reactor facilities; with simple modifications, laboratories can be adapted for this type of analysis and, apart from counting equipment, the apparatus consists of a few pieces of simple glassware.

TABLE 18. Radionuclides of argon

Reaction	Half-life	Energy (MeV)
^{39}K (np) ^{39}Ar	300 years	E_β 0·57
^{40}Ca (nα) ^{37}Ar	35 days	K-capture
^{40}A (nγ) ^{41}Ar	1·8 h	E_β 1·2
^{41}K (np) ^{41}Ar	110 min	β 1·24, 2·55

The main nuclear reactions are given in Table 18 from which it can be seen that radioactive argon isotopes are produced both from argon and from potassium. This allows the simultaneous determination of both potassium and argon from a sample, although the former can also be determined by potassium radionuclides as previously described. Argon is best determined by use of the ^{41}Ar radionuclide which has a characteristic short half-life of 1·8 h; argon produced both by potassium and calcium can be resolved by use of aluminium absorbers.

Analysis can be made by two methods: in one the powdered sample is irradiated and the argon released by heating; in the other the argon is extracted from the sample and then irradiated. In the latter the amount of atmospheric contamination can be measured by using ^{37}Ar formed from ^{36}Ar. The gas evolved from the sample is chemically cleaned by conventional methods utilizing copper oxide and titanium furnaces, and soda-lime and calcium chloride absorption columns. The purified argon is then collected on active charcoal held at −195°C. The final counting of the 1·2 MeV β-rays can either be made in a flow proportional counter, or the argon can be compressed into small containers with end-window mylar films and the activity measured with a conventional end-window Geiger counter.

VII. Dating Geological Material by the Potassium-Argon Method

A. Igneous and Metamorphic Rocks

If an igneous rock has not been subject to a later metamorphic event then the potassium–argon age of a mineral will be that of the time of emplacement provided that argon has not been lost through diffusion. In the case of young intrusions, some time may lapse before the intrusion has cooled down sufficiently so that argon no longer escapes. An emplacement age is one that marks the time at which diffusion rate does not exceed that of radiogenic argon accumulation. Lovering (1955) has suggested a cooling interval of one million years, while Larsen (1945) considers that a cooling interval of seven million years is required for a batholith to cool from 800°C to 200°C. The simple cooling event of an individual intrusion is often complicated by the common sequence of multiple intrusion by which a single intrusion may be heated up several times during the regional intrusive episode.

In cases of complete recrystallization, as a result of a later metamorphic event, previously accumulated argon is completely lost and the metamorphic event is dated. Transition zones in which older ages are only partially lost can exist and have been described by Long, Kulp and Eckelmann (1959).

The dating of minerals assumes that they do not incorporate significant amounts of radiogenic argon at the time of crystallization. Damon and Kulp (1958) showed that beryl, cordierite and tourmaline contained excess radiogenic argon and helium. The excess argon is probably trapped in channels present in the ring structure of these minerals. Hart (1961) used amphibole and pyroxene for dating and for the former was not able to show any evidence of excess radiogenic argon; Hart and Dodd (1962) have shown that pyroxenes can contain excess argon and this was confirmed by Evans (1963) and McDougall and Green (1964); the effect of the excess argon upon the ages of some pyroxenes is shown in Table 19. The excess argon is possibly trapped in crystal imperfections and along low angle grain boundaries; the possibility that the excess argon is a reflection of potassium loss is not considered likely; so far, excess argon has not been observed in micas or feldspars.

The potassium–argon method is of particular value in the case of very young and basic rocks, both of which are difficult if not impossible to date by any other method. Ages of 1·57 and 1·69 million years have been obtained by Curtis, Lipson, Evernden (1956) for an andesite flow and rhyolite plug of Sutter Buttes, California, while McDougall (1963, 1964) has dated whole

TABLE 19. Excess argon in pyroxenes

Rock type	Mineral	% K	Radiogenic ^{40}Ar 10^{-5} ml s.t.p./g	Apparent K–Ar Age (million years)
Amphibolite A.	Hornblende	1·40	7·39	1000
Amphibolite B.	Hornblende	1·30	5·96	900
Pyroxene gneiss	Pyroxene	0·102	0·937	1500
Skarn in amphibolite	Pyroxene	0·00868 0·00859	0·969	10,400

From Hart and Dodd (1962).

rock and biotite from basic rocks from Hawaii. The K–Ar results show that the order of cessation of volcanism occurred along the island chain from north-west to southwest. Most of the exposed volcanoes were built in less than 0·5 million years and all the lavas are late Pliocene and Pleistocene, excepting the Mauna Kuwale trachyte, which is early to middle Pliocene. On the basis of these age measurements, the hypothesis advanced by Wilson (1963) for the origin of the Hawaiian island chain implies that the minimum velocity of convection currents in the mantle is of the order of 10–15 cm/year. These ages provided some hitherto non-existent control to the absolute age of these volcanics and showed that estimates of age based on geomorphological arguments are qualitatively correct.

Ages between 1 million and 100,000 years have been obtained for volcanics by Evernden, Curtis and Kistler (1957), while Koenigswald, Gentner and Lippolt (1961), Leakey, Evernden and Curtis (1961) and Curtis and Evernden (1962) have applied the potassium–argon method to the dating of Zinjanthropus, the lower Pleistocene hominid. With further refinements of technique it will soon be possible to make a direct comparison between the carbon-14 and potassium–argon methods.

Measurements on basic rocks by McDougall (1961, 1963) and Schaeffer, Stoenner and Bassett (1961) have shown that the potassium–argon method can give promising results. Erickson and Kulp (1961) have made a detailed mineral and whole rock age study of the Palisade Sill. The sill was chosen as differentiation has resulted in a number of different petrological phases including metamorphic facies at the contacts. Locally biotite is present in sufficient amounts to be used as a reference age for the other facies of the sill. The sill has not been reheated since the original intrusion and the biotite age of 190±5 million years is in agreement with an Upper Triassic age determined by stratigraphic evidence. The results are given in Table 20 and

TABLE 20. Argon retention in the Palisade Sill

Type	% K	Apparent age	Retentivity %
Biotite	6·41	190±5	—
Whole rock at contact	0·52	202	~100
Fine grain whole rock	0·61	202	~100
Medium grain above centre of sill	0·73	166	~85
Medium grain below centre of sill	0·55	142	~75
Coarse whole rock	0·89	162	~85
Hornfels below sill	1·80	193	~100

From Erickson and Kulp (1961).

show excellent argon retention in the fine grained facies compared to argon loss in the coarser types; in fact this age is used as a reference point in construction of the Time Scale.

B. SEDIMENTS

The dating of sediments by the potassium–argon method has been described by Lipson (1956, 1958), Wasserburg *et al.* (1956), Curtis and Reynolds (1958), Hower, Hurley, Pinson and Fairbairn (1963) and Hurley (1961). In some instances, whole rock samples of sediments may be used for dating, although it is more common to use a separated mineral phase. In both cases it is essential that they retain the radiogenic argon and do not contain fragments of detrital minerals that may contain relict argon. Whole rock samples of metamorphosed sediments such as slates may give acceptable ages provided that these requirements are valid. In addition, apart from true sediments, volcanic ash intercalated between sediments may be used to date the time of sedimentation. Folinsbee, Baadsgaard and Lipson (1961) have dated Upper Cretaceous ash falls by means of sanidine and biotite separated from a bentonite clay. In a special category, potassium rich water soluble minerals found in evaporites have been studied by Stoenner and Zähringer (1958b).

There is great interest in correlating radioactive age measurements with those of the accepted fossil stratigraphic scale. In general, only approximate ages can be determined by means of intrusive rocks, and an accurate measurement requires the dating of geological formations that contain recognized faunas and floras. An authigenic mineral such as glauconite, which has formed at the water–sediment interface and which occurs with fossil remains and is widely distributed in both space and time, has proved to be most suitable. Glauconite, a potassium bearing mica, retains argon sufficiently, although there is the tendency for glauconite ages to be slightly lower than expected. The interlayered structure of glauconite involves mainly 10 Å non-expandable layers and expandable (montmorillonite) layers. The potassium content of glauconite is inversely proportional to the percentage of expandable layers, the latter being greater in younger than older glauconites. The percentage of expandable layers decreases from about 30% in young glauconites to about 10% in those of early Palaeozoic age. This gradual re-orientation with time may result in argon loss and indicate that some of the old glauconites cannot be regarded as truly authigenic, although some argon loss may be related to deep burial. The mineralogy of glauconite and the conditions necessary for its growth in sediments have been described by Galliher (1935), Takahaski and Yagi (1929), Smulikowski (1954), Warshaw (1957), Burst (1958) and Hower (1961). Poleyava, Murina and Kazakov (1961) has described the extensive use of glauconite dating to Russian rocks from the Miocene to Pre-Cambrian age. Ages ranging between 500 and 1300 million years have been obtained for glauconites associated with the pre-Olenellus fauna of the Russian Platform. Apart from such biological remains provided by spores and algae, glauconite is the only suitable mineral available for dating this important part of the time scale.

Methods for determining the age of the source material of sediments have been investigated by Krylov and Silin (1959, 1960), Krylov, Baranovskaya and Lovtsyus (1958) and Krylov, Baranovskaya, Lovtsyus Drozhzhin and Litvina (1958), who have shown that, in the mechanical breakdown of a rock such as a granite, there is initially a loss of radiogenic argon in excess of potassium, but with further breakdown the loss is about equal. The mechanical disintegration of feldspar through the stages coarse-grained–sand–clay–silt does not result in argon loss, while chemical weathering can cause up to 30% of argon to be lost. Krylov has further shown that original potassium–argon ages are retained by the various mechanical breakdown products of a granite from the whole rock to silts; these results are given in Table 21. This approach has also been applied to the dating of silts contained within icebergs. Krylov (1961) obtained ages ranging between 350–690 million years for silt removed from Antarctic icebergs; these results reflect the generally accepted age pattern of Antarctic rocks. In a similar manner Hurley (1963) has shown that acceptable potassium–argon ages can be obtained from the clay fraction separated from palaeontologically dated shales between Ordovician and Recent times, Pelagic sediments (Hurley, Hunt, Pinson and Fairbairn, 1963) from the North Atlantic gave ages between 200–400 million years, the bulk of the potassium is present in illite which is transported to the ocean by the wind and observed age differences with depth are related to transport, wind, source direction and current movement.

TABLE 21. Decomposition of granite ("deluvial" type [1] of weathering, long-distance transport by rivers, and weathering by lake surf)

No.	Rock	Particle size (mm)	K(%)	Ar (cm^3/g × 10^{-5})	Time (million years)
1	Granite (monolith)	—	3·01	4·82	380
2	Granite (monolith)	—	3·43	5·09	355
3	Strongly weathered granite	—	3·56	4·94	335
4	"Deluvium" [1] (gravel)	1–5	3·0	3·08	250
		(Trans: 10–50?)	2·91		
5	"Deluvium" [1] (sand)	3–10	2·43	3·15	265
6	"Deluvium" [1] (fine sand)	0·3–0·5	2·78	3·34	330
7	River sand	0·3–0·5	2·40	4·24	365
8	River pebbles	4–8	3·51	3·85	380
9	Lake sand (at river mouth)	1–2	2·83	5·29	360
10	Lake sand (at river mouth)	1	4·43	4·72	395
11	Lake pebbles	5–8		7·52	400
12	Delta clay	0·01	2·98	4·11	335

[1] Deluvium is debris transported down mountainsides by gravity, rain, or snow.
From Krylov (1961).

Lippolt and Gentner (1963) have dated fossil bones, limestones and fluorite. The concentration of argon was measured with a sensitive mass spectrometer and the potassium content determined by isotope dilution. An acceptable total fossil bone age was obtained, while ages obtained from bone apatite and tooth remains showed considerable argon loss. Assuming that the whole bone age is attributed to calcite replacement, the ages of limestone samples were thus determined. One sample showed an age in excess of that expected and this is attributed to the inclusion of detritus; it would appear possible that pure limestones may be dated, although the retention of argon in calcite is not known. Three other limestones gave similar ages, but these were lower than the accepted age, but approximated to a later period when the limestone was subjected to surface erosion with the development of a Karst type of

topography. The dated fluorites contained between 37 and 67 ppm of potassium; one sample had an age that was expected while the other two had
high ages reflecting the presence of excess argon probably present within the
fluid inclusions.

VIII. Argon Loss from Geological Material

In potassium–argon dating, the production through radioactive decay
of a gaseous end-product raises the problem of the retention of a gas in a rock
or mineral in relation to time and geological processes. The loss of argon is
closely related to the mineral type, and experiment has shown that while
micas are the most suitable material, feldspars, apart from sanidine, in
general show argon loss. Whole rock samples cannot normally be used for
dating purposes, although some fine-grained types can yield acceptable
results. Natural glasses are not suitable as, apart from a high coefficient of
diffusion, devitrification of the glass with time is accompanied by argon loss

In general terms, argon loss can be attributed to three causes.

A. DIFFUSION PROCESSES

The diffusion flow of a substance in a mixture with other substances is
defined as the amount of substance passing perpendicularly through a
reference surface of unit area during unit time. The dimensions of the flow
rate are expressed in cm^2/sec. Diffusion is a process which leads to an equal-
ization of concentration within a single phase, while the laws of diffusion
relate the rate of flow of the diffusing substance with the concentration
gradient responsible for the flow. The coefficient of diffusion need not necessarily be a constant for a given medium, temperature or pressure, and is
usually only approximately true. A study of diffusion in solids, liquids and
gases has been described by Carslaw and Jaeger (1947) and Jost (1952).

Gentner and Trendelenberg (1954), Reynolds (1957), Evernden, Curtis,
Kistler and Obradovich (1960), Baadsgaard, Lipson and Folinsbee (1961),
Amirkhanoff et al. (1961) and Gerling, Morozova and Kurbatov (1961a)
have described diffusion processes in sylvite, feldspar, biotite, phlogopite,
glauconite, amphibole and pyroxene. Some of the various ways in which
diffusion processes occur in nature may be summarized as follows.

(1) Argon loss at room temperature is improbable as the diffusion constants
are very small. Fechtig, Gentner and Kalbitzer (1961) have described the
use of neutron-generated ^{39}Ar in natural silicates to measure the diffusion
characteristics of argon in these substances, and they obtained a value of
$< 5 \times 10^{-24}$ ml/sec for the volume diffusion of argon at room temperature.

(2) Observations of the rate of diffusion of argon in various minerals with
increasing temperature suggest a complex system of argon loss. Sharp knees

the diffusion curves are related to argon loss concurrent with the loss of bsorbed water at 100°C and later the loss of water from the actual lattice at 00–700°C.

. ARGON LOSS ASSOCIATED WITH MINERALS PHASE CHANGES IN THE SOLID

Argon loss occurs along cracks in minerals and also from areas of crystals which boundaries have been created by phase transitions during geological me.

In micas, the radiogenic argon appears to be located in two different ositions, while that present in feldspars appears to be all derived from one 'pe of site. While most feldspars show an argon loss of 10–50%, Baadsgaard al. (1961) has shown that fresh sanidines from volcanic bentonites retain gon sufficiently well to yield reliable ages. In the light of the sanidine results, may be profitable to study early high-temperature homogeneous and later w-temperature exsolved feldspar phases in cogenetic rocks. Assuming no ter argon loss, any observed age differences may be related to the length time required for the magma to cool.

The use of feldspars for dating has been enhanced by the work of Amirk-noff et al. (1961), who heated natural feldspars between 0–700°C in the 'esence of a saturated solution of thallium nitrate. Up to about 500°C the dspar samples lost between 25–30% of potassium by the substitution of allium in the potassium sites. Above about 500°C the amount of thallium bstitution became constant and ages obtained from such feldspars (stable ne) were similar to those obtained from coexisting micas. Prior to the aximum thallium substitution, the feldspar samples gave lower ages, ustrating a typical argon loss pattern. The results of these experiments are ven in Table 22.

TABLE 22. Apparent and true ages of feldspars and micas

Feldspar No.	Age (millions of years) Feldspars		Micas
	Bulk sample	Stable zone	
N319	1720	2020	2010 (phlogopite)
N35	1530	2000	—
N65	1670	1950	—
N823/5	1380	1860	1880 (biotite)
N1/1–5	1360	1575	1560 (biotite)
N6	160	275	285 (phlogopite)

From Amirkhanoff et al. (1961), p. 273.

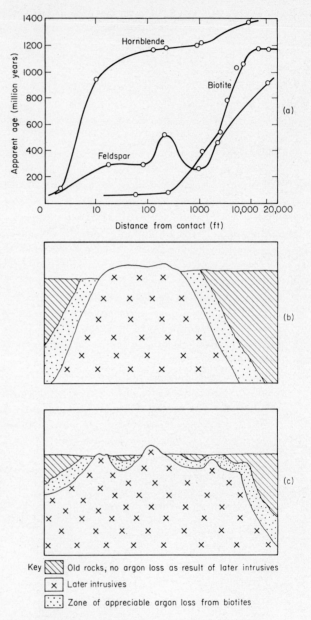

FIG. 16. Argon loss in relation to intrusive contacts and structure. (a) Variation of K–Ar ages as a function of distance from an intrusive contact (Doe and Hart, 1963). (b) and (c) Relation between argon loss and structure.

C. Argon Loss as the Direct Result of an Applied External Stimulus

Foremost in this field is argon loss caused by metamorphism. Most biotites would appear to loose all their argon during a metamorphism and the build-up of radiogenic argon following a metamorphism serves to date this event. Apart from thermal effects, argon is also lost in areas that have been subjected to dynamic forces such as folding and crushing.

In the simple case of a later intrusion emplaced into older rocks, heat produced along the contact zone will result in argon loss from the older material. The magnitude and extent of argon loss will be mainly dependent upon the heating, but also upon the ability of a particular mineral to retain argon and the crystal size. Doe and Hart (1963) have described a Tertiary quartz monzonite intruding Pre-Cambrian gneisses in Colorado, U.S.A.; apparent K–Ar ages as a function of distance from the intrusive contact are shown in Fig. 16. While straightforward argon loss profiles may be found in cases where the contact zone is steep, in others a shallow contact zone will lead to more extensive "aureoles" of argon loss. When erosion has only exposed small isolated areas of a large, mainly hidden, igneous mass, variable argon loss patterns may be observed, as is shown in Fig. 16(c).

Some loss of argon may occur during the crushing and grinding of a mineral, but will often depend upon the amount of lattice disorder and chemical alteration. Mackenzie (1953) has shown that structural changes occur in micas during mechanical grinding. Argon loss is related to structural changes (Gerling et al., 1961b) and not to the radii of the crushed fragment. For micas the bond energy is far too low to keep argon within the lattice, but the ionic bonds of ions surrounding the argon atoms are far too strong to permit their escape. Argon loss occurs when the mineral lattice is broken down, as is the case when structural water is evolved (Sardanov, 1961).

4

RUBIDIUM–STRONTIUM METHOD

I. Introduction

It was first observed by Thomson (1905) and later confirmed by Campbel and Wood (1906) that rubidium is naturally radioactive, but it was not unti 1937 that Hemmendinger and Smythe (1937), Hahn, Strassmann and Wallin̦ (1937) and Mattauch (1937) identified the radioactivity as originating from the rubidium isotope of mass 87. By β-particle emission, rubidium-87 decay to strontium-87, and the discovery of radiogenic ^{87}Sr in Rb-rich Sr-poor minerals, such as lepidolite, led to the use of the ^{87}Sr (daughter)/^{87}Rb (parent ratio for measuring geological age. The natural isotope abundances of rubi dium and strontium are given in Table 23. In the case of a Rb-rich minera containing no common strontium, the age can be directly calculated by determining the total strontium content (i.e. predominantly ^{87}Sr) but, in the more general case in which common strontium is present in moderat₅ amounts, special techniques are required to determine the total amount o₁ radiogenic strontium in the presence of normal ^{87}Sr. The rate of decay o $^{87}Rb \rightarrow ^{87}Sr$ is very slow relative to the age of the earth, and the fundamenta law of radioactivity can be simplified to the form

$$N = N_0^{-\lambda t}$$

where N_0 is the number of ^{87}Rb atoms disintegrated, N is the number of atoms of ^{87}Sr formed and t is the age in years.

$$\therefore \quad t = \frac{N_0 - N}{\lambda N}$$

Rubidium-87 and strontium-87 have the same mass number and therefore

$$t = \frac{\% \text{ radiogenic } ^{87}\text{Sr}}{\% \ ^{87}\text{Rb} \times \lambda}$$

TABLE 23. Isotope abundances for common rubidium and strontium

Isotope	Abundance	Reference
^{87}Rb	27·85	Nier (1950a)
^{85}Rb	72·15	
^{88}Sr	82·56	
^{87}Sr	7·02	Bainbridge and Nier (1950)
^{86}Sr	9·86	
^{84}Sr	0·56	

The initial Rb–Sr dating work was mainly applied to rubidium-rich strontium-poor minerals such as lepidolite or pollucite in which the total strontium was predominantly or solely radiogenic. The total Rb–Sr content was determined by conventional wet chemical analysis, and a mass-spectrographic assay was used to determine the relative amount of radiogenic strontium. An example of the latter assay is given in Fig. 17 in which $SrBr_2$ was used as a source from which Sr ions were produced and registered on a photographic plate (Mattauch, 1938, 1947). This technique is no longer used; instead, strontium isotope abundances are determined by mass spectrometry. The mass spectrum for strontium separated from a rubidium-rich mineral is given in Fig. 18. However, it would appear that in the future the mass spectrograph (Leipziger and Croft, 1964) will become a widely used instrument in the earth sciences. It can be used to measure the abundance of some 70 elements simultaneously for major, minor and trace elements without any large correction required for variations in sample matrix. By replacing the spark source with a thermal ionization source, and photographic plate collection by some form of electrical recording, the conventional mass spectrograph can be modified to be used as a dual purpose machine for the accurate measurement of relative isotope abundances and for the determination of total concentration of an element.

The optical spectrograph has been extensively used (Noll, 1934; Ahrens, 1952) to determine the total Rb and Sr content of suitable minerals, while

Hybbinette (1943) used a preliminary chemical concentration followed by a spectrographic analysis. Ahrens (1948) has determined the Rb/Sr ratio directly by optical spectroscopy using standards having the same matrix composition. The differences in volatility of Rb and Sr in the d.c. arc were overcome by converting all elements in the sample to the volatile fluorides. Conventional spectrographic techniques, however, have been unable to distinguish between the different strontium isotopes.

Strontium-87 alone possesses a magnetic moment, which results in a splitting of the hyperfine structure. A pure sample of ^{87}Sr shows characteristic hyperfine patterns that are not observed for normal strontium. This technique has been used by Heyden and Kopfermann (1938) and more recently

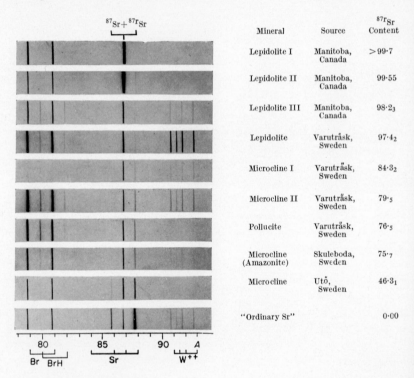

FIG. 17. Mass spectra of $SrBr_2$, obtained from old pegmatite minerals. (Reproduced by courtesy of Prof. J. Mattauch.)

by Eckhoff (1960). The fine atomic structures are not suitable for general use, but by using a large dispersion, high resolution optical spectrograph, Ahrens (1948) was able to resolve ^{88}SrF, ^{87}SrF, ^{86}SrF, in some SrF band spectra. While such techniques showed promise, they were not developed for use in routine analysis and were rapidly surpassed by the higher accuracy and precision provided by the use of stable isotope dilution analysis.

Prior to 1952, reliable age determinations were mainly obtained by U–Pb methods, which restricted the scope of study to ore-bearing rocks. The use of lepidolites by Rb–Sr dating opened up the field somewhat, but it was, in general, still restricted to late-stage pegmatites. Although Goldschmidt (1937) had suggested the use of leucite as a suitable material, it has not been used. Ahrens (1946, 1948), Whiting (1950) and Holyk (1952) have suggested that the common rock-forming mineral, biotite, should in many cases be most

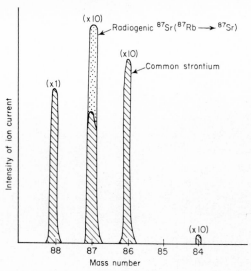

FIG. 18. Common strontium separated from an old rubidium-rich material.

suitable for Rb/Sr dating. Analytical techniques were by now sufficiently advanced for it to be practical to separate and determine small amounts of strontium with the required accuracy. By mass-spectrographic techniques 300 μg of total strontium were required, while in present mass spectrometry 1 μg will suffice for analysis. The remaining problem of determining Rb, Sr normal and Sr radiogenic with sufficient accuracy at trace levels was overcome by Davis and Aldrich (1953) by use of isotope dilution analysis. The use of ion-exchange resins (Aldrich, Herzog, Abelson and Bolton, 1952; Davis and Aldrich, 1953) made the quantitative separation of Rb from Sr possible. Aldrich, Herzog, Holyk, Whiting and Ahrens (1953) determined the isotopic composition of strontium separated from a wide range of source materials and confirmed the earlier suggestion of Ahrens that biotite was a mineral suitable for Rb–Sr dating, provided that large amounts of common strontium were absent. The first biotite ages from Pre-Cambrian rocks were determined by Tomlinson and Das Gupta (1953) and Davis (1954). Following the use of biotite, many other common rock-forming minerals such as musco-vite and K-feldspar were used. For the first time it became possible to obtain mineral ages on a variety of rocks.

On the assumption that a late metamorphism would homogenize the common strontium isotopes and radiogenic strontium present in a rock, the metamorphic event could be dated by using the amount of radiogenic stron-tium formed in a mineral since the metamorphism. In the simple case it is assumed that during homogenization the small amount of radiogenic strontium formed in a mineral prior to the metamorphism is diluted by common strontium of a normal composition.

Homogenization of isotopes is easy to visualize in the case of complete melting, but not in the case of minor recrystallization. It has generally been presumed that radiogenic strontium is loosely held in the crystal lattice, particularly so for micas where it exists in sites previously occupied by rubidium which in turn is at the sites of potassium atoms.

Deuser and Herzog (1963) have presented preliminary evidence, obtained from hydrothermal leaching experiments under pressure-temperature condi-tions, showing that, for pegmatitic biotite and muscovite, common strontium is easily leached. Common strontium is not uniformly distributed throughout the mineral structure and for the most part does not replace potassium. It is also suggested that hydrothermal leaching can be used as a means for re-moving common strontium from samples having an unfavourable Rb/Sr ratio.

Homogenization of strontium isotopes in a closed system is shown in a diagrammatic form in Fig. 19. For practical purposes the closed system is often taken to be a condition that has existed within a hand-specimen of a rock. While Rb and Sr may have completely interchanged within the sample, the sample itself has acted as a closed system and Rb or Sr has been neither lost nor gained during the event responsible for the mixing. While closed chemical systems can be shown to exist down to the scale of geological hand-specimens, geological and geochemical observations are not always compatible with the process. In a simple case, heating together with partial melting would be expected to remove rubidium and potassium relative to strontium. If isotopic homogenization does not occur during the initial phases of re-melting, the mobile phase will be enriched in rubidium, and perhaps radio-genic strontium, while the remainder will be depleted in both. In such a hypothetical case a condition exists such that the mobilized fraction will contain excess ^{87}Sr relative to the depleted fraction, and if this occurs over local distances with a chemically uniform rock, the whole sample, depending where it was collected, can have two different $^{87}Sr/^{86}Sr$ ratios (see p. 93) at the time the event occurred. It is quite common to observe that elements such as K, Na and Fe can migrate over considerable distances during a metamorphic event, or during the final crystallization of a magmatic body. In terms of the age and balance of strontium isotopes, if isotopic homogenization occurred in both phases, the age and initial ratio of a whole rock sample would be meaningful; if not, anomalies should be detected in both the age and initial $^{87}Sr/^{86}Sr$ ratios.

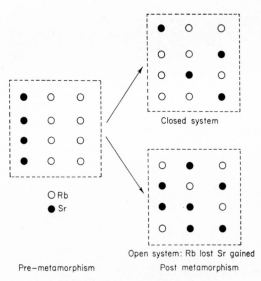

Closed system

○ Rb
● Sr

Open system: Rb lost Sr gained

Pre–metamorphism   Post metamorphism

Fig. 19. Closed and open chemical systems.

II. Decay Constants

The present status of the Rb–Sr method of dating is such that it is now possible to measure the total common rubidium, strontium and radiogenic strontium content of a suitable rock or mineral with both high precision and accuracy. There still remains, however, a little doubt as to the real half-life and decay constant for ^{87}Rb.

Hahn *et al.* (1937) separated strontium from a lepidolite containing a major amount of rubidium; they showed by mass spectrographic analysis that it consisted of 99% radiogenic strontium, and inferred that it had been produced by the radioactive decay of ^{87}Rb. The $^{85}Rb/^{87}Rb$ ratio for rubidium separated from five natural silicates was shown by Brewer (1938b) to be $2·6 \pm 0·01$, while later Nier (1950a) obtained a value of $2·591$, which is currently used in all laboratories. In 1957, Jamieson and Schreiner found differences in the $^{85}Rb/^{87}Rb$ ratio for rubidium separated from some old African lepidolites. Differences might be expected for very old Pre-Cambrian minerals containing small amounts of rubidium. In such a case rubidium will be depleted in ^{87}Rb, but in terms of an age is likely to result in an error of only a few per cent Variations in the $^{85}Rb/^{87}Rb$ ratio have not been supported by the recent work of Shields *et. al.* (1963) who separated rubidium from 27 natural silicates ranging in age from 20 to 2600 million years and obtained a $^{85}Rb/^{87}Rb$ ratio of 2.5995 ± 0.0015. It would appear quite certain that no variations in the atomic abundance ratio of $^{85}Rb/^{87}Rb$ exist in nature and any variations that may exist are less than those accounted for by experimental error.

The determination of the half-life and decay constant for ^{87}Rb has been approached in two ways.

A. Geological Determination

In the early work Hahn *et al.* (1937) obtained a value of $6.5 \pm 0.6 \times 10^{10}$ years for the half-life of rubidium-87 based upon the Sr/Rb ratio of a lepidolite of known age. Using similar material Strassmann and Walling (1938) reported a value of 6.3×10^{10} years. In recent years Aldrich, Wetherill, Davis and Tilton (1956) and Aldrich and Wetherill (1958) have measured the lead–uranium age and $^{87}Rb/^{87}Sr$ ratio for six samples ranging in age from 375 to 2700 million years. Both uranium–lead and rubidium–strontium ages were found to be concordant for a half-life of $5.0 \pm 0.02 \times 10^{10}$ years and this value has been adopted by many laboratories.

B. Physical Determination

A major difficulty inherent in the application of radioactive counting techniques to determine the specific activity of ^{87}Rb and one which has as yet not been satisfactorily overcome, is caused by errors in the measurement of the weak energy tail of the β-spectrum. The determination of the half-life of ^{87}Rb by physical methods falls into three groups.

(1) The half-life has been determined on natural rubidium salts by solid β-counting techniques. Both thin source counting and absorber techniques are handicapped by difficulties in measuring the weak-end part of the β-spectrum.

(2) A possible method of overcoming the absorption of the low energy part of the spectrum would be by putting a rubidium salt into a scintillation medium. Lewis (1952) incorporated rubidium iodide into a thallium activated sodium iodide scintillation crystal. While this technique was virtually free from source absorption and scattering problems, it was still insensitive to the low energy electrons, which would tend to give rise to high results. In addition it is difficult to prepare blank crystals having the same response to the background counting rates. Some of these problems have been overcome

by Flynn and Glendenin (1959) and Glendenin (1961) who incorporated a soluble rubidium salt into a scintillation liquid. The troublesome background effects can be controlled by adding a quencher, and thermionic photo-multiplier noise corrected for by the use of pulse height discrimination. The background counts may also be reduced by refrigeration and by the use of an anti-coincidence circuit. For this work it is essential to check the rubidium salt for cation, anion and radioactive impurities. The total amount of rubidium added to the scintillation liquid is determined by the destruction of the organic liquid followed by a gravimetric analysis of rubidium as the sulphate.

The maximum energy for the β-particles is 275 keV, but the average energy is only 45 keV and the method is only capable of counting energies

TABLE 24. Half-life determinations for ^{87}Rb

Author(s)	Year	Half-life (10^{10} years)
Hahn, Rothanback	(1919)	7·3
Mühlhoff	(1930)	12·0
Orbain	(1931)	4·4
Strassman, Walling	(1938)	6·3
Chaundhury, Sen	(1942)	7·5
Eklund	(1946)	5·8
Haxel, Houtermans, Kemmerich	(1948)	6·0
Kemmerich	(1949)	4·1
Curran, Dixon, Wilson	(1951)	6·2
Curran, Dixon, Wilson	(1952)	6·4
Lewis	(1952)	5·9
Geese-Bähnisch, Huster, Walcher	(1952)	4·8
McGregor, Wiedenbeck	(1952)	6·4
McGregor, Wiedenbeck	(1954)	6·2
Flinta, Eklund	(1952)	6·1
Geese-Bähnisch, Huster	(1954)	4·3
Fritze, Strassmann	(1956)	4·6
Wetherill, Tilton, Davis, Aldrich	(1956)	5·0
Aldrich, Wetherill, Davis, Tilton	(1956)	5·0
Libby	(1957)	4·9
Aldrich, Wetherill	(1958)	4·9
Flynn, Glendenin	(1959)	4·7
Rausch, Schmidt	(1960)	4·7
Ovchinnikowa	(1960)	5·0
Egelkraut, Leutz	(1961)	5·8
Beard, Kelly	(1961)	5·3
McNair, Wilson	(1961)	5·3
Leutz, Wenninger, Ziegler	(1962)	5·80
Fritze, MacMullin	(1964)	4·6

above 150 keV with 100% efficiency, as below this the efficiency is reduced. From a specific β-activity of $55 \cdot 0 \pm 0 \cdot 6$ disintegrations/min per mg, a half-life of $4 \cdot 7 \times 10^{10}$ years and a decay constant of $1 \cdot 475 \pm 0 \cdot 03 \times 10^{-11}$ years^{-1} have been obtained. The major source of error in this technique, amounting to about 5%, is caused by the necessity of extrapolating the integral counting rates, which indicate that the maximum half-life would be $4 \cdot 95 \times 10^{10}$ years.

(3) McNair and Wilson (1961) using exceptionally stable 4π counting techniques have obtained a value of $5 \cdot 25 \pm 0 \cdot 10 \times 10^{10}$ years. Instead of natural Rb they used enriched ^{87}Rb (99%), where the isotopic composition was determined by mass spectrometry. Measurements made on the enriched material were found to agree exactly with those on normal rubidium.

Kulp and Engels (1963) have compared K–Ar and Rb–Sr ages for a large number of samples and show that a half-life of $4 \cdot 70 \pm 0 \cdot 10 \times 10^{10}$ years results in concordance between the two methods. An average deviation of $0 \cdot 6 \pm 0 \cdot 8\%$ for 33 muscovites was found if a half-life of $4 \cdot 7 \times 10^{10}$ years was used, while a net average deviation of $+6\%$ would occur using the $5 \cdot 0 \times 10^{10}$ years value. In a similar manner, out of 80 biotites from igneous rocks 71 showed concordance using a half-life of $4 \cdot 7 \times 10^{10}$ years. Recently, Fritze and McMullin (1964) obtained a tentative value of $4 \cdot 6 \pm 0 \cdot 3 \times 10^{10}$ years by measuring the amount of ^{87}Sr produced in several kilograms of a very pure rubidium salt during a known interval of time. While geological evidence would appear to support a half-life of $4 \cdot 6 \times 10^{10}$ years, the true value is still uncertain, but would certainly lie between $4 \cdot 7$ and $5 \cdot 3 \times 10^{10}$ years. A summary of the various half-lives obtained for ^{87}Rb is given in Table 24.

III. The Determination of Geological Age by the Rb–Sr Method

The Rb–Sr age of a rock or mineral is obtained from the amount of radiogenic ^{87}Sr that has been produced with time by the radioactive decay of ^{87}Rb. Strontium in geological samples can be divided into two types, *common strontium*, and *radiogenic strontium* which accumulates at the sites of rubidium atoms. The relation between both types is given in Fig. 20.

With the exception of a mineral or whole rock containing major amounts of rubidium and only trace amounts of strontium (in which case the ratio ^{87}Sr/^{86}Sr for common strontium can be assumed to be $0 \cdot 712$), two types of sample are necessary. One should contain sufficient ^{87}Sr so that the ^{87}Sr/^{86}Sr ratio can be measured with sufficient accuracy, the other should contain only common strontium, e.g. apatite, calcium feldspar. The acceptable minimum Rb/Sr ratio is about $0 \cdot 5$, but will depend upon the age of the sample.

Rubidium–strontium dating has been applied with much success to metamorphic rocks. Unless other evidence is to the contrary, it is generally assumed that during a metamorphic event there is a complete homogenization of strontium isotopes. In such a case, any ^{87}Sr that may be present is diluted by the common rock strontium and the new ratio $^{87}Sr/^{86}Sr$ will be slightly greater than that prior to the metamorphic event. The time at which the metamorphic event occurred is given by the amount of ^{87}Sr that has subsequently been produced by the decay of ^{87}Rb.

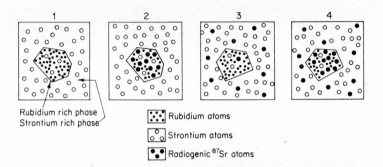

FIG. 20. The relation between common and radiogenic strontium. Sequence 1–4 represent a hypothetical igneous rock consisting of two phases, one rubidium-rich strontium-poor, the other strontium-rich rubidium-poor.

Stage 1. The sample contains rubidium and strontium, but no radiogenic strontium.

Stage 2. After a period of time radiogenic strontium accumulates in the Rb-rich phase.

Stage 3. Represents a stage at which the strontium isotopes were homogenized and the radiogenic strontium is distributed throughout the rock. After this event the common strontium now contains additional ^{87}Sr.

Stage 4. The present day distribution of strontium and rubidium, in which a new quantity of radiogenic strontium has accumulated in the rubidium-rich mineral.

The age of the Rb rich mineral at stage 4 will date the event at stage 3, i.e. radiogenic Sr formed during the period between stage 3 and 4.

A. MINERAL SEPARATION

Before proceeding with the separation of the minerals it is convenient, where possible, to remove a small fragment of a mineral phase for a preliminary semi-quantitative determination of Rb and Sr. This is most conveniently carried out by optical spectroscopy or X-ray fluorescence analysis, and the result will indicate the suitability of the material. If suitable, a representative sample is crushed, a portion retained for whole rock analysis, and the remainder for mineral separation. The actual procedures adopted vary considerably, while a general mineral separation sequence for a granite

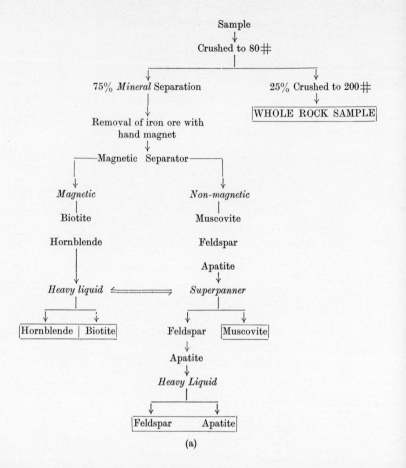

(a)

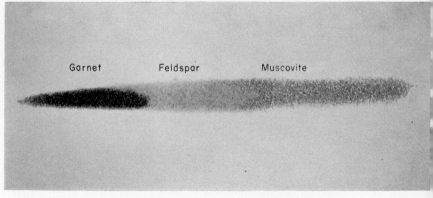

(b)

FIG. 21. (a) Preparation of mineral and whole rock from a granite (b) Separation of feldspar, muscovite and garnet by means of an automatic water panning separator.

is given in Fig. 21(a). An example of the separation of garnet, beryl, feldspar and muscovite by the use of a panning device is given in Fig. 21(b). A review of mineral separation techniques is given by Wager and Brown (1960).

B. Preparation of Standard Tracer

The standard tracer solution which consists of the isotopic enriched isotope can be prepared by either a gravimetric or an isotope dilution method.

1. Gravimetric method

An accurately known weight of the tracer (i.e. $^{87}Rb_2CO_3, ^{86}SrCO_3$) is dissolved in a known volume of dilute acid. The concentration of the tracer is then calculated from the known dilution factor, chemical purity and isotopic

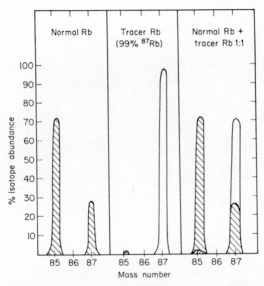

Fig. 22. Determination of rubidium by isotope dilution.

purity. This method of calibration is rarely used as it is subject to errors caused by cation and anion impurity of the tracer. Commercially available rubidium-87 tracer may contain significant amounts of potassium.

2. Isotope dilution method

In this method an approximately known weight of tracer is diluted to an approximately known volume and the concentration is determined by analysis against a solution of normal rubidium and strontium in which the isotopic composition and concentration are accurately known. Unlike the gravimetric method, the chemical purity of the tracer is not important, but only that of the standard solution.

From the preliminary estimation of the rubidium and strontium content of the sample, sufficient tracer is added to the sample such that a favourable ratio can be measured on the mass spectrometer. For the determination of total rubidium and strontium a ratio of 1 between the most abundant normal isotope and the tracer is preferred, i.e. $^{87}Rb/^{85}Rb = 1$, $^{88}Sr/^{86}Sr = 1$. In the case of radiogenic strontium the most favourable ratio is $^{87}Sr/^{86}Sr = 0.712$,

i.e.
$$\frac{^{87}Sr}{^{86}Sr} = \frac{^{87}Sr \text{ normal} + {}^{87}Sr \text{ radiogenic} + {}^{87}Sr \text{ tracer}}{^{86}Sr \text{ normal} + {}^{86}Sr \text{ tracer}}$$

The method of isotope dilution is described in Chapter 2, and an example of the procedures used for rubidium is given in Fig. 22.

C. CHEMICAL SEPARATION

In a mass spectrometer, rubidium ions are produced with greater ease and at a lower filament temperature than those of strontium. In practice this means that, in a mixture of rubidium and strontium, a large ion beam of rubidium is produced relative to that of strontium. The isobaric interference between ^{87}Rb–^{87}Sr necessitates the complete removal of rubidium from

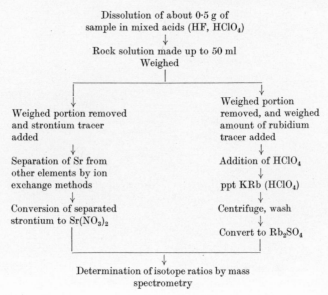

Dissolution of about 0·5 g of
sample in mixed acids (HF, HClO₄)
↓
Rock solution made up to 50 ml
Weighed

Weighed portion removed
and strontium tracer
added
↓
Separation of Sr from
other elements by ion
exchange methods
↓
Conversion of separated
strontium to Sr(NO₃)₂

Weighed portion
removed, and weighed
amount of rubidium
tracer added
↓
Addition of HClO₄
↓
ppt KRb (HClO₄)
↓
Centrifuge, wash
↓
Convert to Rb₂SO₄

↓
Determination of isotope ratios by mass
spectrometry

FIG. 23. General scheme for the separation of rubidium from strontium in rocks and minerals.

strontium when measuring strontium isotopes. The presence of strontium in a rubidium analysis is of no consequence, as the running temperature for the emission of strontium ions is about 2·3 A compared to 1·3 A for rubidium.

Chemical procedures vary considerably in different laboratories, but a general scheme for the separation of rubidium from strontium is given in Fig. 23. While organic ion-exchange resins are generally used for the separation of rubidium and strontium, inorganic exchange materials such as zirconium phosphate or molybdate appear to show promise. Apart from exceptionally clean separation, these methods require very small column volumes of exchange media and use small volumes of eluting reagents. Both these factors become important in reducing blank values, apart from making the separation more complete and rapid.

D. MASS SPECTROMETRIC DETERMINATION

The separated rubidium or strontium is placed on a heated tantalum filament, and the filament placed into the source of a mass spectrometer. The temperature of the filament is increased until a steady ion beam is produced. The relevant mass range is then repeatedly scanned until some 25–30 scans have been obtained. The required isotopic ratios and the standard deviation of the run are then calculated. Errors can arise during the emission of ions either by contamination within the source region so that strontium and rubidium ions begin to be emitted from various parts of the source, or by isotopic fractionation of the isotopes during the analysis. Fractionation from the filament occurs both for rubidium and strontium and is in part at least controlled by the physical state of the filament and chemical purity of the sample. The magnitude of fractionation varies in both direction and magnitude with time, and is distinct from regular discrimination caused by ion optics and effects of fringing fields, which can be considered to be constant with time.

In a normal fractionation sequence the heavier isotope will increase with time. Jäger (1962) has reported a 1% change for two mass units for strontium and 1·5% for two mass units for rubidium. Fractionation of strontium isotopes can be corrected in isotope dilution analysis by the use of a mixed ^{84}Sr–^{86}Sr tracer in which the $^{88}Sr/^{84}Sr$ and $^{88}Sr/^{86}Sr$ ratios are used to correct for fractionation. This correction is essential when dealing with samples containing very small amounts of total strontium.

When the isotopic composition of common strontium is being determined, the observed $^{87}Sr/^{86}Sr$ ratio is corrected by use of the $^{86}Sr/^{88}Sr$ ratio which is considered to be invariant in nature. The $^{86}Sr/^{88}Sr$ ratio is accepted as 0·1194 and half the difference between the measured and true $^{86}Sr/^{88}Sr$ ratio is used to correct the observed $^{87}Sr/^{86}Sr$ ratio. The total Sr, radiogenic Sr and Rb are then calculated from the basic isotope dilution equation.

The approximate age of the sample can then be obtained according to the equation $t = \log(1 + D/P)$,

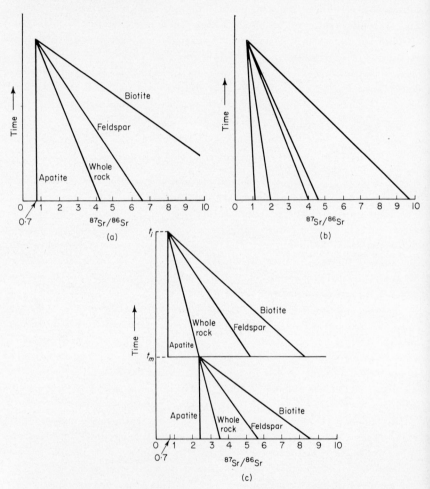

Fig. 24. Graphical illustration of the isotopic composition of strontium by the Compston-Jeffrey method. (a) All phases have remained closed chemical systems with respect to Rb,Sr since crystallization and cooling of the rock. (b) Five whole rock samples from one intrusion having different Rb/Sr ratios remaining as closed chemical systems with respect to Rb,Sr since crystallization and cooling. (c) Initial crystallization of rock at t_i followed by metamorphism or melting at a later date t_m resulting in homogenization of strontium isotopes, while the whole rock remained remained a closed chemical system with respect to Rb,Sr. Note increase in initial $^{87}Sr/^{86}Sr$ ratio from t_i to t_m reflecting growth of ^{87}Sr in a closed system. Sequence of growth lines can be repeated for several metamorphisms; the whole rock line may remain linear, although gain or loss of Rb or Sr at a particular metamorphism may occur.

$$t = \frac{{}^{87}\text{Sr radiogenic}}{\lambda \ {}^{87}\text{Rb}} \quad \begin{array}{l}\text{(daughter)}\\[4pt]\text{(parent)}\end{array}$$

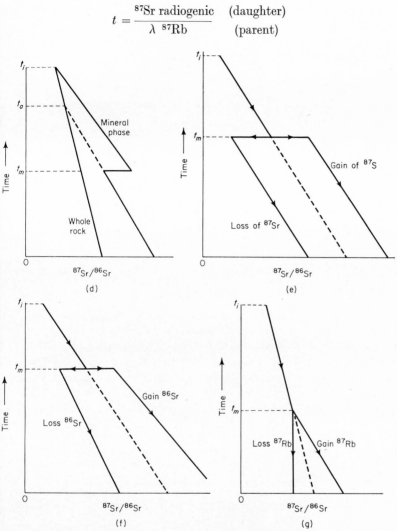

(d) Partial homogenization of Sr at t_m. Growth lines for mineral phase A and whole rock intersect at an apparent age t_a between t_i and t_m. (e) Loss or gain of ^{87}Sr at time of metamorphism t_m for a mineral growth line. (f) Loss or gain of ^{86}Sr at time of metamorphism t_m. (g) Gain or loss of ^{87}Rb (i.e. total rubidium) at time of metamorphism t_m.

IV. The Application of Theoretical Methods to Rb–Sr Dating

During the early period of the strontium–rubidium method, it was assumed that the isotopic composition of the common strontium was the same as that found in sea-water, i.e. ^{87}Sr$/{}^{86}$Sr $= 0.712$. As the amount of common strontium was generally small and the amount of radiogenic ^{87}Sr large, the isotopic composition of the common strontium was not critical. However, as the method was gradually extended to a large range of geological problems, it soon became apparent that it was necessary to be able to determine very small amounts of radiogenic ^{87}Sr, either because the common strontium content of a sample was high, leading to unfavourably low Rb/Sr ratios, or the sample was young, in which case only a small amount of ^{87}Sr would be present.

It now became vital to know the isotopic composition of total strontium, expressed as the ratio ^{87}Sr/^{86}Sr, which is equivalent to the setting of the isotopic clock when it began to measure time. Recently, the works of Compston and Jeffery (1960, 1961), Schreiner (1958) and Compston, Jeffery and Riley (1960) in Australia and Allsop (1961), Hales (1961) and Nicolaysen (1961) of the Bernard Price Institute (B.P.I.) in South Africa, have provided theoretical models which, when applied to geological problems, have marked a major advance in Rb–Sr dating. This type of approach was first suggested by Holmes (1932).

A. THE COMPSTON-JEFFERY METHOD

The age of a rock or mineral is defined as the length of time that has passed since it became a closed chemical system with respect to Rb and Sr. Under these conditions the age is given by the amount of radiogenic ^{87}Sr that has formed during time t (t = age of sample) and can be determined since

$$^{87*}\text{Sr}^{(1)} = {}^{86}\text{Sr} \left[(^{87}\text{Sr}/^{86}\text{Sr})_{\text{now}\,(2)} - (^{87}\text{Sr}/^{86}\text{Sr})_{\text{initial}\,(3)} \right],$$

(1) ^{87*}Sr = radiogenic component.
(2) now = present day.
(3) initial = ratio at time of emplacement, i.e. the last time that homogenization of the strontium isotopes occurred.

Thus

$$T = {}^{86}\text{Sr}/\lambda^{87}\text{Rb} \left[(^{87}\text{Sr}/^{86}\text{Sr})_{\text{now}} - (^{87}\text{Sr}/^{86}\text{Sr})_{\text{initial}} \right]. \tag{16}$$

As mass spectrometers only measure ratios, the amount of ^{87}Sr, which is a variable, is compared to the amount of the invariant isotope ^{86}Sr.

Equation (16) is that of a straight line with a gradient of ^{86}Sr/λ^{87}Rb and with ^{87}Sr/^{86}Sr initial as an independent variable. Using this approach, a series of minerals or a whole rock having different Rb/Sr ratios will have individual lines when plotted on linear graph paper, with time plotted on the abscissa and ^{87}Sr/^{86}Sr on the ordinate. The slope of the line will be steep for a rubidium-rich phase such as biotite, and shallow for a strontium-rich phase such as apatite. The slope of the whole rock line will be intermediate between the two extremes as it is a weighted mixture of both phases. When the strontium isotopes were last homogenized, and if since this event they remained in a closed system, all growth lines should converge to a point marking the age of the sample, while the intercept on the $^{87}Sr/^{86}Sr$ axis will give the initial $^{87}Sr/^{86}Sr$ ratio at the time of homogenization. The point of intersection is one at which the $^{87}Sr/^{86}Sr$ ratio for different mineral phases is the same as that for the whole rock.

In Fig. 24(a) the Compston-Jeffery model is given for a sample that was homogenized with respect to strontium at a time t_0. This may indicate the crystallization of a magma or a last metamorphism.

If a rock, which has undergone a single metamorphism at some time after its initial formation, remains a closed system with respect to Rb and Sr, and if the second event results in the homogenization of the strontium isotopes, the increments of radiogenic strontium formed in the different phases are diluted by the greater abundance of common strontium present in the whole rock. This is best shown by the Compston-Jeffery plot given in Fig. 24(c) indicating the complete homogenization of Sr isotopes at a time of metamorphism. In a closed system the slope of the whole rock growth line is unaltered and the intercept on the abscissa will give the age of the first event, while the convergence of the mineral lines at t_m will date the metamorphic event. After this, the slope of the new growth lines would not necessarily be the same as before, since during the metamorphism, there would undoubtedly have been a partition of Rb and Sr among the newly formed minerals. The difference between the initial $^{87}Sr/^{86}Sr$ and $^{87}Sr/^{86}Sr$ at the time of metamorphism will depend upon the time that has passed before the metamorphic event.

A case of partial homogenization of strontium at a time of metamorphism is given in Fig. 24(d). In this instance, the growth line of the Rb-rich mineral intersects the total rock growth line at an apparent age t_a between the initial age t_1 and the later metamorphism t_m. In such cases the actual age t_1 is best sought in areas where the effect of the metamorphism at t_m is slight. Failing this, if an $^{87}Sr/^{86}Sr$ of 0·71 is assumed and if the real value is lower, an anomalously high age is obtained and, if it is high, then the age for t_m will be too low. If the slope of the growth curve is steep, the calculated age is not critically dependent upon the initial value, but if it is shallow then the intercept with the whole rock time will be critical, and unrealistic old ages may be obtained.

Problems involving partial homogenization are found in metamorphic terrains and also possibly in the case of mantled gneiss domes, although in the latter case complete homogenization of the strontium isotopes may also occur.

Although strontium isotope studies indicate that closed systems are common, there are cases where loss or gain of Rb, Sr common or radiogenic ^{87}Sr has occurred. The effects in such cases upon the growth lines are given in Fig. 24(e–g).

The Compston-Jeffery method for depicting a metamorphic event illustrates that in a closed system it should be possible to find minerals with a ^{87}Sr/^{86}Sr ratio significantly greater than 0·71 that have been affected by a metamorphic event. Compston and Jeffery (1959) showed that biotite, microcline, apatite and epidote separated from the Boya granite, W. Australia, were indeed enriched in radiogenic strontium. If the original ^{87}Sr/^{86}Sr ratio were assumed to be 0·71, a range of ages between 650 and 2430 million years would be obtained. By inserting an initial ratio of 0·82, that had been obtained by graphical analyses on the different mineral phases, by the Compston-Jeffery method all the ages were reduced to 520 million years. This is interpreted as showing that the granite was emplaced about 2430 million years ago, but 520 million years ago was subjected to a metamorphic event that resulted in the homogenization of the strontium isotopes, and in a new starting ^{87}Sr/^{86}Sr ratio of 0·82.

The Compston-Jeffery method uses growth lines, a point on which is determined by the ^{87}Sr/^{86}Sr ratio and the age of the sample. The actual age is determined by the intercept of two minerals or the whole rock and at least one mineral phase. The intercept of the growth lines will give the time at which the ^{87}Sr/^{86}Sr in the different phases were the same. In terms of the age, it is from this point that radiogenic strontium starts to accumulate in the individual mineral phases.

B. THE BERNARD PRICE INSTITUTE (B.P.I.) METHOD

The B.P.I. method is mathematically equivalent to the Compston-Jeffery method and, in a graphical analysis in which the present day ^{87}Sr/^{86}Sr ratio is plotted against the present-day ^{87}Rb/^{86}Sr ratio, samples of the same age all lie along a straight line called an isochron.

For a rock containing both initial and radiogenic strontium, the total amount of ^{87}Sr is represented by:

$$(^{87}\text{Sr})_{\text{today}} = (^{87}\text{Sr})_{\text{initial}} + (^{87*}\text{Sr})_{\text{radiogenic}}$$

$$= (^{87}\text{Sr})_{\text{initial}} + {}^{87}\text{Rb}_{\text{today}} \ (e^{\lambda t} - 1)$$

Referring the measured quantities to ^{86}Sr we have:

$$\left(\frac{^{87}\text{Sr}}{^{86}\text{Sr}}\right)_{\text{today}} = \left(\frac{^{87}\text{Sr}}{^{86}\text{Sr}}\right)_{\text{initial}} + \frac{^{87}\text{Rb}(e^{\lambda t} - 1)}{^{86}\text{Sr}}$$

The foregoing equation is that of a straight line of the form $y = mx + c$, where

$$y = \left(\frac{^{87}Sr}{^{86}Sr}\right)_{today} ; \quad x = \frac{^{87}Rb}{^{86}Sr} ; \quad m = e^{\lambda t} - 1 ; \quad c = \left(\frac{^{87}Sr}{^{86}Sr}\right)_{initial}$$

An example of an isochron plot is given in Fig. 25. The intercept of the isochron on the y-axis corresponds to a hypothetical rock with zero Rb, and gives the initial $^{87}Sr/^{86}Sr$ ratio. In the case of two co-existing samples having differing rubidium contents, the sample with the greater amount will produce more ^{87}Sr during an interval of time and consequently will lie further along the isochron than the sample with less rubidium. For a B.P.I. plot to be valid for samples of coeval age the following conditions must be fulfilled.

1. All samples must have the same initial $^{87}Sr/^{86}Sr$ ratio.
2. They must all have the same age.
3. The rock must have acted as a closed system.

These assumptions are usually valid for a rock that has crystallized from a magma, but are not always true for metamorphic or sedimentary rocks in which the initial $^{87}Sr/^{86}Sr$ value will depend upon the source material. The

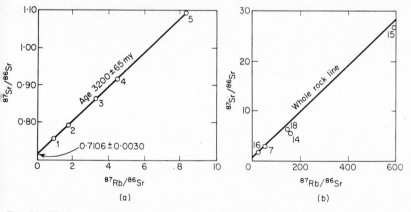

FIG. 25. Rb–Sr age measurements from old granite, Central Transvaal. (a) B.P.I. Plot for whole rock samples. (b) B.P.I. Plot for separated-mineral samples enriched in radiogenic strontium. (After Allsop, 1961.)

model can be applied to minerals, but if they have been metamorphosed the $^{87}Sr/^{86}Sr$ ratios at the time of the metamorphism may vary depending upon the ability of the mineral to retain its radiogenic strontium. A commonly found order of retention is muscovite > feldspar > biotite. For a single whole

TABLE 25. Rb–Sr Age measurements from old granites, Central Transvaal (after Allsop, 1961)

Sample number	Sample	Total rock			
		$\dfrac{^{87}Sr}{^{86}Sr}$ atomic	^{86}Sr (ppm)	^{87}Rb (ppm)	$\dfrac{^{87}Rb}{^{86}Sr}$ atomic
1	Witkoppen	0·753±0·006	32·16±0·32	30·39±0·45	0·934±0·018
2	Corlett Drive	0·789±0·003	26·68±0·10	46·66±0·45	1·729±0·017
3	Halfway-House	0·862±0·010	17·34±0·07	58·25±0·60	3·321±0·037
4	Bryanston	0·915±0·003	13·67±0·21	62·07±0·60	4·488±0·082
5	Honeydew	1·089±0·006	12·08±0·20	101·70±0·50	8·321±0·143

TABLE 25—*contd.*

Sample number	Source	Mineral	$\frac{^{87}Sr}{^{86}Sr}$ atomic	^{86}Sr (ppm)	$^{87*}Sr$ (ppm)	^{87}Rb (ppm)	Apparent age (million years)
6	Halfway-House granite	Feldspar	0·868±0·015	15·66±0·08	2·58±0·08	59·10±1·10	3060±100
7		Biotite	2·664±0·027	5·44±0·03	10·75±0·15	327·5±1·6	2310±40
8		Chlorite	1·040±0·030	3·27±0·02	1·09±0·10	26·50±0·60	2890±280
9		Muscovite A	0·905±0·010	52·61±0·26	10·36±0·54	159·4±2·1	4540±230
10		Muscovite B	0·950±0·010	41·1±1·0	9·95±0·42	182·9±1·9	3820±160
11		Apatite	0·736±0·005	14·09±0·32	0·37±0·07	1·64±0·12	
12		Epidote	0·806±0·005	150·7±1·4	14·47±0·77	9·5±0·9	
13	Witkoppen granite	Feldspar	0·818±0·012	15·99±0·17	1·74±0·19	47·25±0·50	2580±290
14		Biotite	5·489±0·027	1·26±0·01	6·08±0·02	202·4±0·6	2120±10
15	Halfway-House pegmatite	Biotite	26·23±0·25	0·86±0·02	22·13±0·22	514·4±2·5	3010±30
16		Feldspar	1·838±0·015	6·61±0·06	7·54±0·10	176·4±1·5	3000±40
17	Witkoppen pegmatite	Feldspar	0·970±0·005	12·61±0·06	3·32±0·06	88·86±0·51	2620±60
18	Corlett Drive pegmatite	Feldspar	6·60±0·05	2·77±0·04	16·48±0·12	413·8±4·2	2800±30

rock sample it is necessary to assume that the initial $^{87}Sr/^{86}Sr$ ratio was about 0·71 to obtain the intercept on the y-axis; in doing so this fixes the slope of the line. If two whole rocks are available with differing Rb/Sr ratios, it is possible to define both the age and initial $^{87}Sr/^{86}Sr$ ratio. For quantitative studies the method requires the use of at least two whole rocks, but it is an advantage to deal with several in order that standard statistical analysis can be applied. This type of analysis was first used by Schreiner (1958), and its potential value was illustrated by Hales (1961) and Allsop (1961). The latter author analysed five whole rocks and thirteen minerals from an old granite of the Central Transvaal, South Africa (Table 25). The whole rock samples (Fig. 25 (a), (b)) lie along a single isochron with an age of 3200 ± 65 million years (using $\lambda = 1\cdot39 \times 10^{-11}$ year^{-1}) and an initial $^{87}Sr/^{86}Sr$ ratio of $0\cdot7006\ \pm0\cdot003$; the mineral ages (Fig. 25 (a),(b)) were scattered above and below the 3200 million year isochron. This suggests that they had been affected by a later metamorphic event. If the later metamorphism had not been imposed upon the rocks, then apart from loss of radiogenic strontium by diffusion, all the mineral ages would also fall along the 3200 million year isochron.

Both the Compston-Jeffery and B.P.I. methods have resulted in major advances in Rb–Sr dating, and also have opened up new and rewarding approaches to fundamental geological problems by means of strontium isotope studies. As more results become published there is increasing evidence that it is essential to maintain a close relationship between strontium isotope studies and conventional geological studies.

The following items serve to illustrate some features that may often have to be taken into account in terms of either interpreting age patterns or reducing the error on the Rb–Sr ages.

(1) If a basement granite is affected by a later metamorphism, parts of the granite may remain unaffected while others will be subject to complete homogenization of strontium isotopes during the metamorphic event. In such a case it is possible to obtain a transition zone between the true basement age and the later metamorphic event,

The presence of a transition zone may, on the other hand, signify several separate metamorphic events of differing ages. If the metamorphism should affect a sedimentary series and convert the members to their metamorphic equivalents, there is the additional problem of inherent radiogenic ^{87}Sr.

(2) In metamorphic terrains, in particular, the location and field setting of the sample often has to be known in great detail for Rb/Sr ages to become meaningful. This point can be illustrated by taking a hypothetical case of a metamorphism produced from depth in an orogenic belt. It is assumed that the central axis of the belt has been subjected to high grade metamorphism together with local melting; passing outwards, the effects of metamorphism decrease. In such a case it is reasonable to suggest that the central area will

record ages of the metamorphism, while in the surrounding area the magnitude of metamorphism will decrease and the effect of different initial $^{87}Sr/^{86}Sr$ ratios will become apparent. Concordant Rb–Sr ages would therefore be obtained in zones surrounding the central mass, while a traverse across the

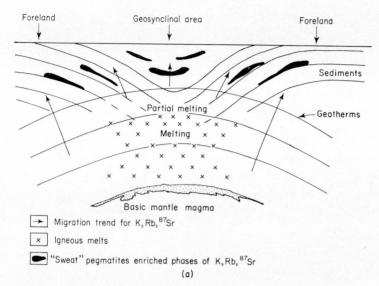

Foreland   Geosynclinal area   Foreland

Sediments

Partial melting

Geotherms

Melting

Basic mantle magma

→   Migration trend for K, Rb, ^{87}Sr

×   Igneous melts

⬛   "Sweat" pegmatites enriched phases of K, Rb, ^{87}Sr

(a)

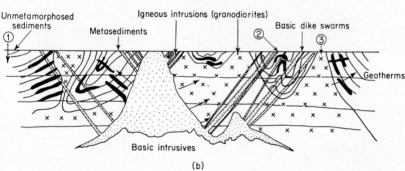

Unmetamorphosed sediments ①

Metasediments

Igneous intrusions (granodiorites)

Basic dike swarms ③

②

Geotherms

Basic intrusives

(b)

Fig. 26. Migration of K, Rb and ^{87}Sr during orogenesis. (a) Mobilization of alkali elements ahead of regional melting. The rise and fall of the geotherms will control the length of time ^{87}Sr can migrate or diffuse. (b) Early "sweat" pegmatites folded and cut by late "sweat" pegmatites at culmination of orogenic event. Local remorphism and contact metamorphism adjacent to acid and basic intrusions results in migration of K, Rb and ^{87}Sr.

whole mass will result in discordant mineral ages. In addition, the waning stages of a metamorphism may often be associated with some form of hydrothermal activity that permeates the area from below. This can lead to the removal of either Sr normal, Sr radiogenic, or Rb from substantial areas or along narrow belts. Pegmatite formed in metamorphic terrains can be of the "sweat" type and may contain excess radiogenic strontium that has been removed from the surrounding or deeper rocks, as illustrated in Fig. 26.

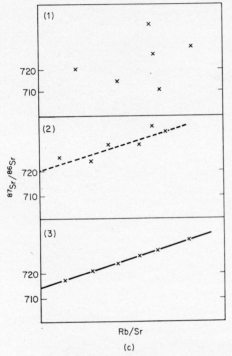

Original sediments, unmetamorphosed. The scatter of points reflects differences in Rb and Sr content of sediment type, e.g. shales, arkoses, limestones, sandstones.

Replacement, and sweat pegmatites, dominantly a low temperature alkalirich liquid that has leached Na, K, Rb, ^{87}Sr in passage through the sediments. Points approximate to an isochron assuming a process of homogenization has taken place in the migrating liquids.

Large mixed igneous magmas formed during the orogeny. Points all lie along a single isochron, showing complete mixing and a closed chemical system.

FIG. 26 (contd.) (c) ^{87}Sr/^{86}Sr against Rb/Sr for whole rock samples. Taken from locations 1, 2 and 3 on part (b).

(3) If an intrusion took a substantial time to cool, the cooling interval as a time factor plus the possible difference of Rb and Sr distribution during this period could be reflected by diffusion ages. In general, these differences would be within the experimental error, but might become significant in the case of very young intrusives.

(4) From geochemical studies it is accepted that weathering can cause the migration of many elements. Zartman (1963), from a study of the Llano Uplift, Central Texas, obtained from fresh material concordant biotite ages of 1020–1030 million years, while weathered biotite gave concordant K–Ar/Rb–Sr ages 3–10% lower, and biotite from a pegmatite exposed in a quarry for twenty years gave Rb–Sr ages 10–40% lower. Fluorite from the pegmatite was encrusted with carbonate that contained an excess of radiogenic strontium distributed by ground water. An adjacent foliated granite

gave concordant K–Ar ages but the Rb–Sr ages were not only discordant, but different from each other by about 10%. These differences are attributed to differing Rb contents and either the migration of ^{87}Sr from the rock or the addition of Rb to it.

(5) In a thorough and very careful study of Rb–Sr ages from Alpine rocks, Jäger, Niggli and Baethe (1963) correlated age with metamorphism caused by burial. A study of separated biotites showed that several mineralogically different biotites were present each having a different age related to a separate metamorphic event. It is quite common for a metamorphic event detected by Rb–Sr dating to leave no obvious imprint on the hand specimen or in a thin section.

V. Rb–Sr Dating of Sedimentary Rocks

If sediments could be dated by the Rb–Sr method it would be possible to produce an accurate geological time scale in relation to fossil remains, but at present the dating of whole rock metasediments and sediments is still very much in its infancy. Radiogenic ^{87}Sr present in sediments is distributed between that contained in detrital grains and that formed in authigenic minerals developed in the sediments at the time of deposition or soon after. Coarse-grained sediments would tend to retain their own radiogenic strontium, while extremely fine-grained sediments such as mudstones, clays, shales and slates would be subject to a more homogeneous distribution of strontium isotopes. As the sediments are produced from a source rock, chemical transport may lead to the selective removal of strontium or rubidium during transportation. Under such conditions the isotopic abundance and concentration of strontium in the source material and the resultant sediments may be very different. While strontium may tend to be coprecipitated with calcium carbonate, rubidium is geochemically more mobile and, in the sediments, is likely to be fixed by absorption on fine-grained particles or trapped in the pore spaces of sediments (Witney, 1962).

The problem and effect of inherited radiogenic strontium in sedimentary rocks has been described by Compston and Pidgeon (1962). Recently, Witney and Hurley (1964) have shown that, while the tendency must be towards higher ages, if the provenance of the sediments is young, or where there has been a selective enrichment of Rb relative to Sr during weathering, transportation and sedimentation, this maximum age may be close to the real age. In the case of the Hamilton shales (Fig. 27, Middle Devonian, New York and Pennsylvania), the low age of the source materials is mainly responsible for the relatively small difference between the measured age and estimated true age.

Before graphical procedures can be applied to dating total sediments, they must have had a uniform $^{87}Sr/^{86}Sr$ ratio at the time of deposition (within the limits of experimental error); also they must have remained closed chemical systems with respect to Rb and Sr from the time of their deposition to the present day.

The age of sediments can often be obtained by more precise methods by dating (a) ash falls intercalated in sediments (e.g. sanidine, biotite), or (b) authigenic minerals present in sediments.

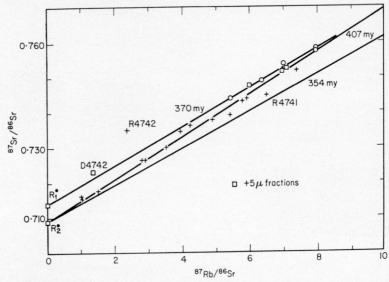

Fig. 27. Limits of isochrons from the Hamilton shales (Witney and Hurley, 1964) (my = million years.)

For such minerals to be of general use they must fulfil the following requirements.

(i) The mineral should be formed at the same time as or soon after the deposition of the enclosing rocks.

(ii) The mineral should be of common and widespread occurrence and be found throughout a stratigraphic succession of known relation to faunal assemblages.

(iii) The mineral should be of the Rb-rich Sr-poor type, while another mineral containing no rubidium (such as apatite or gypsum) would be an advantage in determining the original $^{87}Sr/^{86}Sr$ of the sediments.

(iv) A clear distinction must be made between a pure authigenic mineral and a detrital mineral that has authigenic overgrowths.

(v) Whole rock samples of sediments can be used provided that the portion of detrital minerals containing inherited strontium is absent or can be allowed for in the age calculation.

The most commonly used authigenic mineral in Rb–Sr dating is glauconite (Hurley, 1961; Poleyava *et al.*, 1961). While glauconite forms through authigenic processes, the mineral undergoes slow composition purification with time, and such examples are not truly authigenic and may account for the lower ages obtained for some Lower Palaeozoic glauconites.

Strontium isotope studies have introduced terms such as complete homogenization of isotopes, closed chemical systems and whole rock samples. In many cases, the interpretation of strontium isotopes can be made and is valid only in terms of these facts or assumptions. However, while these features are easily accepted in age studies, fundamental geological and geochemical theory may seriously question their validity.

VI. The Significance of the Ratio $^{87}Sr/^{86}Sr$ in Petrogenetic Problems

The application of isotope ratios to study petrogenetic problems was first suggested by Holmes (1932), who supposed on the evidence available at that time that ^{41}K decayed to ^{41}Ca; it is now known that ^{40}K is the radioactive isotope that decays to ^{40}Ar (see p. 48). Holmes's suggestion that the proportion of Ca to ^{41}Ca would increase something like twenty-seven times as fast in granites as in basalts, because of the lower abundance in the former than the latter, was the forerunner of the now accepted graphical approach to Rb–Sr dating. Holmes also proposed that the accumulation of stable daughter isotopes, the abundance of which could be correlated to a particular rock type such as basalt or granite, could be used to study fundamental theories of petrogenesis. This has now been accomplished by studies of the isotopic composition of strontium and lead.

Before discussing the application of strontium isotopes to problems of petrogenesis, it is first necessary to elaborate certain assumptions, or describe some possible processes that are compatible with the observed data.

A. THE RELATION OF THE EARTH TO METEORITES

The average chemical composition of the earth, as a whole or of the mantle, is very similar to that of chondrite meteorites. Assuming that the earth and meteorites were formed from the same system, it is probable, apart from perhaps local differences, that both had the identical isotopic composition for common strontium. Any variations observed in the $^{87}Sr/^{86}Sr$ ratio are due solely to original differences in the relative abundances of Rb and Sr and any changes that have occurred are due only to the radioactive decay of ^{87}Rb.

Meteorites having different Rb/Sr ratios are amenable to isochron studies and show that the common initial ^{87}Sr/^{86}Sr ratio was about 0·698, while ages obtained from the slope of the isochron show that chemical differentiation in chondrites and achondrites occurred during the interval $4 \cdot 3 – 4 \cdot 7 \times 10^9$ years ($\lambda = 1 \cdot 47 \times 10^{-11}$ year^{-1}). The isotopic composition of strontium in stone meteorites is given in Table 26 and a corresponding isochron plot in Fig. 28. Meteorite ages by Schumacher (1956), Herzog and Pinson (1956), Webster (1958), Gast (1960a, 1962) and Anders (1962) are given in Table 26.

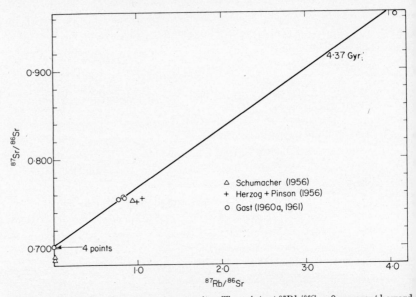

FIG. 28. ^{87}Rb–^{87}Sr isochron for stone meteorites. The points at ^{87}Rb/^{86}Sr$\sim$0 represent howardites and eucrites; the point at $\sim$4 represents Beardsley; the points at $\sim$0·8 represent the remaining chondrites. A least-squares fit of Gast's points gives an isochron with a slope corresponding to $t = 4 \cdot 37 \times 10^9$ years (with a Rb half-life of 46×10^{10} years; Flynn and Glendenin, 1959).

The enrichment of Rb in meteorites can be compared to normal igneous differentiation, while Urey (1951) and Gast (1960b) have suggested that some of the ages may reflect the selective loss of Rb. In the case of the Beardsley chondrite, Gast (1960b, 1961) has shown that much of the "excess" K, Rb Cs is readily leachable by water, suggesting the presence of a condensed vapour phase.

By identifying the earth with meteorites the initial ^{87}Sr/^{86}Sr ratio for terrestrial material should be 0·698. The most direct comparison that can be made between meteorites and terrestrial material is that of the world-wide basaltic magma, which for present purposes is regarded as being representative of mantle material.

TABLE 26. Isotopic composition of strontium in meteorites (Anders, 1962)

Meteorite	Class	Sr(ppm)	Rb(ppm)	$^{87}Sr/^{86}Sr$
Beardsley II	Cg	10·4	14·25	0·960 [a]
Forest City	Ccb	10·2	2·71	0·755 [a]
		12·0	3·91	0·754 [b]
		9·8	3·50	0·756 [c]
Holbrook	Cik	12·8	2·22	0·739 [a]
Homestead	Cgb	10·6	3·60	0·753 [c]
Modoc	Cwa	10·1	2·96	0·757 [a]
Richardton	Cca	10·1	2·85	0·757 [a]
Bustee	Bu			0·685 [b]
Moore County	Eu	79·5	0·16	0·701 [a]
Nuevo Laredo	Ho	84·4	0·37	0·703 [a]
Pasamonte	Ho	92·7	0·22	0·701 [a]
				≤0·703 [c]
Pasamonte (white)		94·7	0·50	0·689 [b]
Pasamonte (grey)		89·5	0·66	0·685 [b]
Sioux Country	Ho	68·8	0·23	0·702 [a]

[a] Gast (1960b, 1961).
[b] Schumacher (1956).
[c] Herzog and Pinson (1956).

B. $^{87}Sr/^{86}Sr$ RATIO IN BASALTS

Gast (1960b), Faure and Hurley (1963), Murthy and Steuber (1963), Hedge and Walthall (1963) and Hamilton (1963) have shown that, with our present techniques of measurement, the initial $^{87}Sr/^{86}Sr$ ratio for oceanic basalts is in the range 0·702–0·705. Continental basalts show a range of 0·702–0·711, suggesting that they are slightly enriched in ^{87}Sr. While these differences are small, the histogram given in Fig. 29 indicates the significant differences that are observed.

Much attention has been focused on the $^{87}Sr/^{86}Sr$ ratios from Hawaiian rocks, as these are well removed from continental areas. The results obtained by Faure (1963), Powell, Faure and Hurley (1964), Lessing and Cantanzaro (1964) and Hamilton (1964) given in Tables 27 and 28 show remarkable agreement among the $^{87}Sr/^{86}Sr$ ratios, although minor differences have been observed. Powell et al. (1964) have tentatively concluded that the $^{87}Sr/^{86}Sr$ ratios of the undersaturated rocks are significantly lower than those of the more silica-rich samples, supporting the hypothesis that Hawaiian lavas are derived from more than one parent magma. Lessing, Decker and Reynolds (1963) and

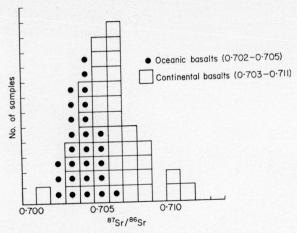

FIG. 29. Initial $^{87}Sr/^{86}Sr$ ratios for oceanic and continental basalts (Faure and Hurley, 1963; Hedge and Walthall, 1963; Hamilton, 1963).

TABLE 27. $^{87}Sr/^{86}Sr$ ratio in Hawaiian volcanics

Rock Type	Locality	$^{87}Sr/^{86}Sr$
Olivine basalt	Kilauea, 1894, Hawaii	0·7036 [a]
Olivine basalt	Kilauea, Hawaii	0·7043 [a]
Basalt glass	Kilauea, 1921, Hawaii	0·7043 [a]
Basalt	Maui	0·7017 [a]
Trachyte	Hualalai, Hawaii	0·7059
Trachyte	Hualalai, Hawaii	0·7072
Trachyte	Hualalai, Hawaii	0·7054
Trachyte glass	Hualalai, Hawaii	0·7052
Hawaiite	Mauna Kea, Hawaii	0·7048
Hawaiite	Mauna Kea, Hawaii	0·7043
Hawaiite	Mauna Kea, Hawaii	0·7046
Ankaramite	Halaeakala, Hawaii	0·7033
Nepheline basalt	Honolulu, Oahu	0·7044
Mugearite	Kohala, Hawaii	0·7052
Linosaite	Kohala, Hawaii	0·7052
Basalt flow	Kapoha, Hawaii, 1960	0·7046
Basalt cinder	Iki Kilauea, Hawaii, 1959	0·7056
Tuff	Diamond Head, Oahu	0·7038

[a] From Faure and Hurley (1963). Remaining values are from Lessing and Cantanzar (1964).

Lessing and Catanzaro (1964) have observed a consistent inverse relation between the $^{87}Sr/^{86}Sr$ ratio and the K/Rb ratio. They suggest that this may be caused by assimilation of marine sediments which has subsequently influenced the petrogenesis of the Hawaiian magma.

TABLE 28. Comparative study of the $^{87}Sr/^{86}Sr$ ratios in some Hawaiian rocks

Rock type	Locality	$^{87}Sr/^{86}Sr_{corr.}$[†] A [3]	B [4]	Difference [5] between A, B
Quartz diabase	A [1] Pabolo Quarry,	0·7047	—	—
	B [2] Honolulu Oahu	0·7041	—	—
Trachyte	Puu Anahulu, Hawaii	0·7048	0·7043	0·0005
Tholeiite	1881 Flow Mauna Loa, Hawaii	0·7048	0·7041	0·0007
Hawaiite	S. Popo Gulch Mauna Kea, Hawaii	0·7040	0·7043	0·0003
Ankaramite	Popo Gulch, Mauna Kea	0·7044	—	—
Alkali olivine basalt	Keauhou Beach, Hawaii	0·7036	0·7040	0·0004
Nepheline basalt	Honolulu Volcanic Series, Oahu	0·7023	0·7031	0·0008
Melilite nepheline basalt	Moiliili Quarry, Honolulu Oahu	0·7030	0·7030	0·0001
	Average	0·7040	0·7038	0·0002

† Normalized to 0·1194.
[1] Main quartz diabase.
[2] Narrow $\frac{1}{2}$ in. vein of more acid diabase.
[3] Analyst E. I. Hamilton, Oxford (1965).
[4] Analyst J. L. Powell, M.I.T. (1963).
  Both sets of samples were presented independently by Professor G. A. Macdonald.
[5] The differences between values in columns A and B are probably related to the use of different mass spectrometers.
  From Anders (1962).

C. $^{87}Sr/^{86}Sr$ RATIO FOR SEA-WATER

The present day $^{87}Sr/^{86}Sr$ ratio for sea-water would appear to lie between 0·709 and 0·712, and has been taken as representing an average for crustal material. When significant differences are observed in present-day sea-water they can be equated to run-off from the continental areas. However, such variations can be expected to be obliterated after complete mixing in the deep ocean basins although it is possible that slight differences may be preserved.

Although no direct sampling of sea-water can be made throughout geological time, the best approximation is afforded by thick limestone deposits. Generally, limestones are precipitated from clear water masses and the direct contribution of continental detritus will be small. Hedge and Walthall (1963) have shown a distinct decrease in the ratio $^{87}Sr/^{86}Sr$ with increase in geological age from 0·7081 to 0·7004, but the number of samples analysed was few. The lowest value from the Pre-Cambrian Bulawayan limestone, which contains primitive organic remains, has also been reported by Gast (1962) and Hamilton and Dean (1963). This limestone occurs in a vast thickness of calcareous and ferruginous sediments associated with basic volcanic rocks, and basic and ultrabasic intrusions. The sedimentary sequence could be interpreted as the breakdown of basic rocks, which themselves would be expected to have low $^{87}Sr/^{86}Sr$ ratios. Also, the sedimentary cycle would lead to an early separation of rubidium from strontium in an environment low in rubidium and high in strontium. The nature of rocks forming shore-facies around basins of limestone deposition may be approached by a study of detrital grains in the limestones. Caution must be adopted in interpreting the data from limestone with particular reference to the type of continental rocks that surround the ancient seas.

VII. Crustal and Mantle Growth Trends Related to Strontium Isotope Studies

Studies of the distribution of radiogenic strontium by Gast (1961) showed that the $^{87}Sr/^{86}Sr$ ratio of basalts ranged from 0·704 to 0·712. Similar measurements made on strontium separated from sea-water, shells growing in sea-water and limestones of different geological ages showed a fairly constant ratio of between 0·711 and 0·712. Unlike the later work of Hedge and Walthall (1963), no variations were observed in the ratio for limestones of early Pre-Cambrian to Ordovician age. Additional calculated initial $^{87}Sr/^{86}Sr$ ratios for a limited number of granites were very similar to the ratios found in modern sea-water and basalts. Gast's observations form the basis for more recent papers describing the significance of strontium isotopes in rocks and minerals.

In a classic paper, Hurley, Hughes, Faure, Fairbairn and Pinson (1962) proposed a model in which rubidium and strontium are used as natural tracers to describe the history of differentiation of the continental sial. Terms used by Hurley are defined as follows.

Basement. The upper 5–10 km of continental crystalline rocks, excluding the thin covering of sediments.

Sial. Material with average composition of the basement. The sial differs from basalts or mantle in containing larger concentrations of alkalies, alumina and silica.

Sub-sialic source region. The source, at some time or other, of sial by differentiation with enrichment in rubidium, other alkalies, silica and alumina, etc.

Primary age. The time that has passed since the initial differentiation gave rise to a Rb/Sr ratio greater than the sub-sialic region and consequently a greater $^{87}Sr/^{86}Sr$ ratio.

Geologic age. The time at which a geological unit was formed, time of emplacement of igneous rock, of metamorphism, or deposition of a sedimentary rock. " Geological " instead of " geologic " is used in this volume.

The enrichment of Rb relative to Sr during differentiation of the source material is related to the development of sial. Faure and Hurley (1963) estimated that the Rb/Sr ratio in the average sial is at least three, and probably five, times greater than that in the subsialic source regions. Using a $^{87}Sr/^{86}Sr$ ratio of 0·708 for the source region of basalts, Hurley and his coworkers calculated the $^{87}Sr/^{87}Rb$ ratio for acid rocks from published data of rock ages. The results of this work, given in Fig. 30(a), clearly indicate that the $^{87}Sr/^{87}Rb$ ratio of an acid igneous rock is proportional to its geological age. Equating this increase in the Rb/Sr ratio with magmatic differentiation also shows that the sial is continuously generated from sub-sialic regions throughout geological time, and the primary age of the basement is only slightly different from the geological age. The slope of the line in Fig. 30(a) corresponds to the decay constant for ^{87}Rb. If a value of 0·708 approximates to the present $^{87}Sr/^{86}Sr$ ratio in the source regions, it does not permit any extensive prehistory of the sial with rubidium. In practice this means that, during an orogenic event, relatively large amounts of freshly differentiated sial are brought up from depth. As new mountain belts will contain strontium having a lower $^{87}Sr/^{86}Sr$ ratio, the derived sediments will have only slightly higher values. The average $^{87}Sr/^{86}Sr$ of clastic sediments formed by erosion of newly formed mountain chains is not high (0·706–0·71) (Witney and Hurley, 1963) and is in fact similar to modern sea-water (0·709–0·71). A further dilution step is added by strontium from volcanic sources. From the published Rb–Sr age data Hurley *et al.* (1962) calculated the $^{87}Sr/^{86}Sr$ ratio at the time of formation for eighty-two granitic rocks according to the equation

$$^{87}Sr/^{86}Sr_{initial} = {}^{87}Sr/^{86}Sr_{today} - {}^{87}Rb \, (e^{\lambda t}-1)/^{86}Sr$$

The mean value of between 0·7070–0·710 for all these measurements shows no preferred variations with time, as shown in Fig. 30(b).

The average Rb/Sr ratio for average sial is about 0·25; ancient sial with an initial $^{87}Sr/^{86}Sr$ ratio of about 0·706 would increase to 0·720–0·730 in 1–2 billion

years. As the observed sialic $^{87}Sr/^{86}Sr$ ratios are low, sediments formed soon
after an orogenic event could be recycled into the new sial, but the results do
not support the idea of reworked ancient sial.

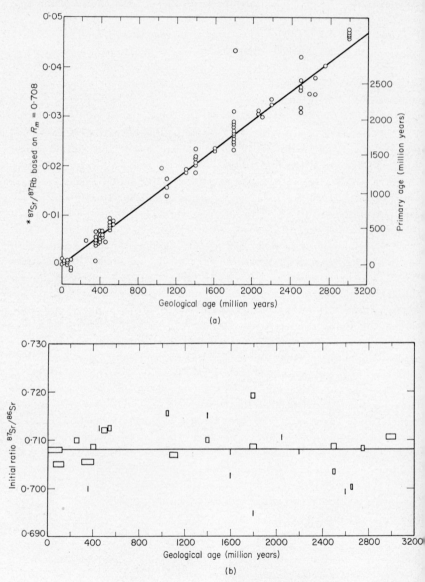

FIG. 30. (a) Variation with geological age of radiogenic $^{87}Sr/^{87}Rb$ in whole rocks. (b) Initial
ratio $^{87}Sr/^{86}Sr$ in sialic rocks of various geologic ages. Weight of measurements indicated by
width of block.

While Hurley et al. (1962) state that the model remains to be proved by more data, it forms a major advance in the application of strontium isotopes to understanding geological problems.

Although Hurley et al. (1963) do not observe any systematic variation in the initial $^{87}Sr/^{86}Sr$ ratios with geological time, Hedge and Walthall (1963) have suggested that, for mafic rocks at least, the ratio tends to lessen with increasing age. To account for the differences between the observed and calculated $^{87}Sr/^{86}Sr$ ratios for the growth of radiogenic strontium in the sial, Hurley et al. (1963) have suggested that a primeval crust has been continuously added to by the influx of new material having a low $^{87}Sr/^{86}Sr$ ratio, from the mantle. Further, this dilution process has been enhanced by the reworking of sialic material, thus diminishing the effect of any local increase in radiogenic strontium by mixing with common strontium. As an alternative, variation of the initial $^{87}Sr/^{86}Sr$ ratio in sialic rocks may indicate that they originated deep within the crust or in the mantle in a region in which the Rb/Sr ratio approximates to that of basalt, but where the strontium is more radiogenic than in the source regions of oceanic basalts. The necessary conditions can also be fulfilled if the crust is layered with respect to radiogenic strontium and its Rb/Sr ratio. In such a case, variations in the initial $^{87}Sr/^{86}Sr$ ratio would be related to depth of origin, although for many there exists the possibility of contamination by addition of radiogenic-rich material.

The significance of current theories about the origin of the sial, as interpreted by means of strontium isotopes, can be fully assessed only when more analyses are available. While the general pattern of $^{87}Sr/^{86}Sr$ ratios in terrestrial material is probably known, further analyses will undoubtedly illuminate the finer points necessary for a more complete understanding of sialic evolution.

VIII. Examples of the Use of the Ratio $^{87}Sr/^{86}Sr$ in Petrogenesis

A. COMAGMATIC RELATIONSHIP OF DIFFERENTIATED IGNEOUS ROCKS

A problem common to many igneous rocks is whether or not a suite of rocks forming a single intrusion, or dispersed throughout a well defined petrogenetic province, have in fact been formed from a single parental magma. If all the rocks have been formed from a single homogeneous magma, then whole rock development lines for each rock type must converge to a point when plotted backwards in time. This point represents the time of differentiation and the $^{87}Sr/^{86}Sr$ ratio of the original magma at that time. Faure (1963) has illustrated this type of analysis in a study of various alkaline rocks of the well known Monteregian Hills, Quebec. The isochron for the (Fig. 31) nordmarkite (quartz-bearing syenite), the more basic vamaskite, essexite, tinguaite, and related carbonatites, clearly indicates that, at about 124 million years ago, a common magma existed, and the alkaline rocks have been formed through a process of magmatic differentiation.

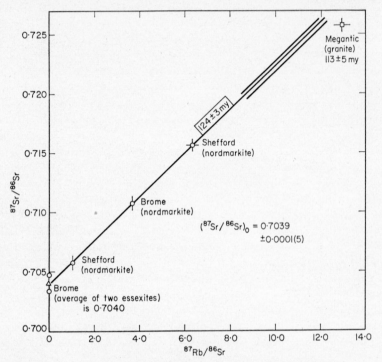

FIG. 31. Whole-rock isochron for nordmarkite and essexite from Brome and Shefford. (From Fairbairn, Faure, Pinson, Hurley and Powell, 1963.)

Basic igneous rocks or provinces are often associated with either minor or major amounts of granitic material. When the latter is present in small amounts by volume it can be explained by simple fractional crystallization; in other cases the volume of acid material is so large as to preclude such an origin unless there exists at depth a vast volume of equivalent basic rocks.

If a sequence of rocks is in fact derived by fractionation from a single magma, then any observed variations in the ratio $^{87}Sr/^{86}Sr$ are related to the growth of ^{87}Sr by the radioactive decay of ^{87}Rb. After correcting for the growth of ^{87}Sr with time, the $^{87}Sr/^{86}Sr$ ratio for a basic rock and petrologically related acid differentiates should be the same. Natural fractionation of strontium isotopes has not been observed. The problem of whether or not granites present in basaltic processes have been formed by differentiation from basalts is particularly amenable to strontium isotope studies. Moorbath and Bell (1965) have measured representative rock-types from the classical Tertiary igneous suite of Skye, northwest Scotland. The initial $^{87}Sr/^{86}Sr$ ratio for twelve basalts, gabbros and peridotites is 0.7058 ± 0.0010 compared to 0.7124 ± 0.0015 for the associated granite. It is concluded that the granitic rocks have been derived from a source with a significantly higher Rb/Sr ratio than those of the basaltic rocks. The most obvious sources are the adjacent Pre-Cambrian Lewisian rocks which form the underlying basement to this area. A study of the well documented differentiated Tertiary Skaergaard intrusion (Wager and Deer, 1939) by Hamilton (1963) has indicated that in this instance the acid-end differentiates do contain excess radiogenic strontium that cannot be ascribed to growth from rubidium *in situ*. The $^{87}Sr/^{86}Sr$ ratios for the Skaergaard intrusion are given in Table 29. In this instance the late stage granophyres have probably assimilated varying amounts of Pre-Cambrian gneiss that forms the country rock. While the strontium isotopes clearly show that these rocks are contaminated, a proportion of the extreme differentiate may have been produced by crystallization differenti-ation. Contamination of some of the Skaergaard acid granophyres by the complete or partial assimilation of country rock is supported by studies of the $^{18}O/^{16}O$ isotope ratios by Taylor and Epstein (1963).

While the use of $^{87}Sr/^{86}Sr$ ratios would appear to be ideal to investigate co-existing and comagmatic basic and acid rocks, it is essential that the $^{87}Sr/^{86}Sr$ and Rb/Sr ratios of the contaminating material is greater than that of the material it contaminated if different $^{87}Sr/^{86}Sr$ ratios are to be of value. In the case of the Skaergaard intrusion the present day $^{87}Sr/^{86}Sr$ ratio of the surrounding basement granite is not regarded as being sufficiently different from that of the Skaergaard basic rocks, and the simple inclusion of average country rock cannot explain in this instance the high ratios found in some of the acid granophyres. In some instances it is possible that partial or selective remobilization of a rock or Rb-rich mineral will provide a highly enriched source of radiogenic strontium.

B. ORIGIN OF ALKALINE ROCKS

Faure (1963) has shown that the initial $^{87}Sr/^{86}Sr$ ratio of the alkaline rocks of the Monteregian Hills is 0.7047 indicating that they have been formed in a crustal environment having a low Rb/Sr ratio similar to that of basalt.

In addition, an initial $^{87}Sr/^{86}Sr$ ratio of 0.704 (Hamilton, unpublished data) has been obtained for the Tertiary alkaline intrusion of Kangerdlussaq, E. Greenland, described by Wager and Deer (1939). It remains to be shown whether or not these values are common to all alkaline rocks. In particular, strontium isotope studies may prove to be useful in examining the relation between, miaskites and agpaites; also their genetic relation to the commonly associated alkaline granites, common to many alkaline intrusions and associated with the more abundant, undersaturated rock types.

TABLE 29. Average initial $^{87}Sr/^{86}Sr$ ratios for the Tertiary Skaergaard intrusion, E. Greenland

Rock type	No. anal.	Height above sea level (m)	$^{88}Sr/^{86}Sr$	$^{87}Sr/^{86}Sr$ [a]	$^{87}Sr/^{86}Sr_{corr.}$ [b]
Marginal gabbros	6	—	0·1191	0·7076	0·7063
Layered series					
Ferrogabbros	6	1924–2450	0·1187	0·7084	0·7065
Hortonolite ferrogabbro	4	1800	0·1196	0·7075	0·7079
Middle gabbro	2	1180	0·1198	0·7075	0·7084
Hypersthene olivine { gabbro {	5	280–825	0·1191	0·7054	0·7044
	1	110	0·1206	0·7028	0·7063
Average layered series	—		0·1195	0·7069	0·7066
Upper border group					
Melanogranophyres	3	—	0·1191	0·7089	0·7071
Transitional granophyres	2	—	0·1192	0·7110	0·7104
Tinden granophyre sill					
Altered acid granophyre	3	—	0·1203	0·7281	0·7303
Fresh acid granophyre	2	—	0·1200	0·7135	0·7141
Melanocratic acid granophyre	1	—	0·1200	0·7094	0·7094

[a] Observed $^{87}Sr/^{86}Sr$ ratio.
[b] Corrected by adjusting the measured $^{88}Sr/^{86}Sr$ ratio to 0·1194 and the $^{87}Sr/^{86}Sr$ by one half this amount. Also corrected for growth of ^{87}Sr in 50 million years.

In a similar manner the $^{87}Sr/^{86}Sr$ ratio of a number carbonatites of differing geological age are characterized by very low and constant $^{87}Sr/^{86}Sr$ ratios of about 0·706 (Powell and Hurley, 1963; Hamilton and Deans, 1963), as shown in Table 30. In Fig. 32 the $^{87}Sr/^{86}Sr$ for carbonatites are compared with the observed ranges found in basalts, sedimentary limestones and marine water. Limestones have in general $^{87}Sr/^{86}Sr$ ratios of 0·712 and consequently carbonatites cannot be formed by the remobilization of limestones, provided that there is not present at depth a limestone having a $^{87}Sr/^{86}Sr$ ratio of about 0·706 or less. $^{87}Sr/^{86}Sr$ ratio of carbonatites indicates that they have been derived from depth, which is in keeping with their mode of intrusion. It is equally apparent that very old Pre-Cambrian limestones can have low $^{87}Sr/^{86}Sr$ ratios which would be found in deeper parts of the crust (see p. 111).

At present it can be concluded that the strontium isotope of alkaline rocks indicates that they have been formed at depth in a low Rb–Sr environment. While the $^{87}Sr/^{86}Sr$ ratios suggest a relationship between alkaline rocks and basalts or deeper ultrabasic rocks such as peridotite, the association of

TABLE 30. The ratio $^{87}Sr/^{86}Sr$ in Carbonatites

Locality	$^{87}Sr/^{86}Sr$ [†]	Locality	$^{87}Sr/^{86}Sr$ [†]
Africa		*Africa continued*	
Busumbu, Uganda	0·7062 [a]	Rangwa, Kenya	0·7072 [a]
Chigwakwalu, Nyasaland	0·7080 [a]	Shawa, S. Rhodesia	0·7064 [a]
Chilwa, Nyasaland	0·7071 [a]	Spitzkop, S. Africa	0·7058 [a]
	0·7044 [b]	Sukulu, Uganda	0·7056 [a]
Galapo, Tanganyika	0·7072 [b]	Tororo, Uganda	0·7056 [a]
Glenover, S. Africa	0·7068 [a]	*Canada*	
Kaluwe, N. Rhodesia	0·7051 [a]	Lake Nipissing, Ontario	0·7039 [c]
Kangankunde, Nyasaland	0·7046 [a]	Nemegosenda Lake, Ontario	0·7043 [c]
	0·7074 [b]	Oka, Quebec	0·7062 [a]
	0·7088 [b]	Seabrook Lake, Ontario	0·7045 [c]
	0·7066 [b]	*United States*	
Loolekop, S. Africa	0·7081 [a]	Iron Hill, Colorado	0·7076 [a]
Mirma, Kenya	0·7074 [a]	Magnet Cove, Montana	0·7076 [a]
	0·7067 [b]	Mt. Pass, California	0·7074 [a]
Muambe, Mozambique	0·7052 [b]	Rocky Boy, Montana	0·7087 [a]
Nkombwa, N. Rhodesia	0·7065 [a]	*Europe*	
Oldolnyo Lengai, Tanganyika	0·7062 [b]	Alno, Sweden	0·7052 [a]
	0·7041 [c]	Fen, Norway	0·7051 [a]
Panda Hill, Tanganyika	0·7034 [b]	Stjernøy, Norway	0·7036 [c]
Premier Mine, S. Africa	0·7058 [a]		

[†] Normalized to 0·1194.

[a] Powell, Hurley and Fairbairn (1962). M.I.T. Report, NYO-3943, p. 16.

[b] Hamilton and Deans (1963). *Nature* **198**, No. 4882, 776.

[c] Powell and Hurley (1963). M.I.T. Report, NYO-10, 517, p. 52.

(The differences between a, c and b are probably related to the use of different mass spectrometers.)

5*

alkaline rocks and trace elements characteristic of sialic material suggests that alkaline rocks in general may be derived from deep areas of the sial having a low Rb/Sr ratio, together with the addition of some material from more basic underlying zones.

While the suggestion of a layered sial with respect to its Rb/Sr ratio may answer some problems, attention must be paid to the mineral assemblages present at depth. The pressure–temperature conditions at the crust–mantle

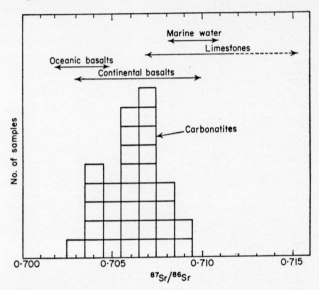

Fig. 32. $^{87}Sr/^{86}Sr$ ratios for carbonatites (see Table 30) compared with those for basalts, sedimentary limestones and marine water. (Note that the range given for sedimentary limestones excludes that of the Bulawayan limestone, p. 111.)

boundary approximate to those of the granulite facies of metamorphism. As a rock passes through the various stages of progressive metamorphism, the tendency will be for the new mineral phases to be depleted in rubidium. Heier (1964) has pointed out that low Rb/Sr ratios have been observed in rock of high granulite facies. The Rb/Sr ratio for these granulites is about 0·1 compared to 0·25 obtained by Faure and Hurley (1963) for near-surface parts of the continental crust.

CHAPTER 6

RHENIUM–OSMIUM METHOD

I. Introduction

Naturally occurring rhenium consists of two isotopes, ^{185}Re and ^{187}Re, with respective abundances of 37·07% and 62·93%. Of the two, ^{187}Re is radioactive and decays by the emission of a β-particle to the stable daughter isotope ^{187}Os. Osmium consists of ^{184}Os (0·018%), ^{186}Os (1·59%), ^{187}Os (1·64%), ^{188}Os (13·3%), ^{189}Os (16·1%), ^{190}Os (26·4%) and ^{192}Os (41·0%); none of these isotopes is radioactive. The geological age relation of ^{187}Re→^{187}Os is similar to that for the decay ^{87}Rb→^{87}Sr, but restrictions related to the natural occurrence of both rhenium and osmium, coupled with a lack of detailed knowledge as to the distribution of these elements and their abundances in most igneous rocks, have limited its usefulness.

Some uncertainty exists as to the half-life of ^{187}Re. Direct counting measurements are handicapped by the difficulty in measuring the very low decay energy of less than 8 keV. The early measurements of Naldrett and Libby (1948) indicated a value of 4×10^{12} years; a later estimate of $6·2 \times 10^{10}$ years was advocated by Herr and Merz (1958) while, more recently, Hirt, Tilton, Herr and Hoffmeister (1963) have obtained a value of $4·3 \pm 0·5 \times 10^{10}$ years based upon Re and Os measurement on minerals of known geological age.

II. Geochemistry

In nature, rhenium does not occur in the native form or as a compound represented by a distinct mineral phase. The detailed geochemistry of rhenium has as yet not been studied in any great detail, apart from the work of Noddack and Noddack (1931), while Goldschmidt (1929, 1954) estimated a crustal abundance of 0·001 ppm. The main economic source of the World's rhenium supply is obtained as a by-product in the refining of molybdenum from molybdenite.

Extensive diadochy exists between quadrivalent molybdenum and rhenium, which have ionic radii of 0·74 Å and 0·72 Å, respectively. In addition, the ionic radii of niobium, 0·69 Å, and tantalum, 0·68 Å, are similar to that of rhenium, which accounts for the presence of rhenium in niobium and tantalum minerals. In molybdenite, rhenium is probably present as ReS_2, which is isostructural with MoS_2. Apart from molybdenite, rhenium is enriched in such minerals as gadolinite, rare earth minerals, columbite, tantalite, oxides of manganese, and in copper sulphide ores (Mansfield, copper schists: Feit, 1930, 1933).

The concentration of rhenium in molybdenite is very variable (Hirt *et al.*, 1963; Badalor, Basitova and Godunova, 1962; Karamyan, 1962) and may be related to rhenium rich and rhenium poor provinces. While the latter may control the general level of rhenium present in an area of molybdenum mineralization, late stage and secondary alteration processes are in particular responsible for the depletion of rhenium in secondary minerals. Moracherskii and Nachaera (1960) have described a secondary ferromolybdenite (Fe_2 $(MoO_4)_3 7HO$) containing 0·5 ppm Re, while the parent molybdenite contained 20 ppm Re. The separation of Mo–Re is further enhanced by the much lower solubility of Ca,Sr,Ba–Mo compounds compared to the very soluble Ca,Sr, Ba–Re compounds. Recently, Poplavko, Marchakova and Zak (1962) have described the occurrence of a rhenium-rich sulphide mineral, dzhezkazganite, the existence of which was shown by micro-probe analysis. It contains 40–50% of rhenium and occurs in disseminated galena–chalcocite–bornite ores.

Geochemical data for osmium is very sparse; it has an estimated crustal abundance of about 1×10^{-7} %. Osmium is practically confined to an association with the platinum group of minerals common to ultrabasic rocks, where it occurs in an alloyed form with iridium.

While osmium can be determined by colorimetric techniques using a thiourea complex, the generally low abundances of both osmium and rhenium are best determined by neutron activation analysis (Herr, Hoffmeister and Langhoff, 1960; Herr, Hintenberger and Voshage, 1954; Leddicotte, 1961b). The nuclear reactions and energies of the resultant isotopes are given in Table 31.

TABLE 31. Nuclear properties of rhenium and osmium

Nuclear reaction	Half-life	Energies (MeV)
$^{185}Re\,(n\gamma)\,^{186}Re$	88·9 h	β^- 0·934, 1·072 γ 0·1372, 0·627, 0·764
$^{187}Re\,(n\gamma)\,^{188}Re$	16·7 h	β^- 1·96, 2·12 γ 0·1551
	18·7 min	I.T. γ 0·0635, 0·105
$^{190}Os\,(n\gamma)\,^{191}Os$	14 h 16 days	I.T. γ 0·074 β^- 0·143
$^{192}Os\,(n\gamma)\,^{193}Os$	1·3 days	β^- 1·13

Isotopic enrichment of ^{187}Os is determined by gas-source mass spectrometry using OsO_4. Hirt *et al.* (1963) have measured 2×10^{-7} g of radiogenic osmium with an accuracy of about 10%.

III. Re–Os Method in Geochronology

The geological age relations are similar to those used in Rb–Sr dating, namely:

$$\frac{^{187}Os \text{ radiogenic}}{^{187}Re} = e^{\lambda t - 1}$$

From the experimental determinations of present day ratios, the ^{187}Os/^{186}Os ratios in the past and also the age t, are calculated. In common with the Rb–Sr method, it is assumed that (a) the law of radioactivity is valid, (b) the Re/Os ratio has been changed only by the decay of ^{187}Re to ^{187}Os and (c) initially the isotopic composition of osmium was the same.

The ^{187}Os/^{186}Os ratio varies between 0·88 and 1·09 in terrestrial materials and between 1 and 1·43 in meteorites. A plot of the measured ^{187}Os/^{186}Os against the ^{187}Re/^{186}Os ratios results in an isochron in which the slope corresponds to the age of the sample, and the intercept on the ^{187}Os/^{186}Os axis is the initial ^{187}Os/^{186}Os ratio.

Applying this approach to meteorites, Hirt *et al.* (1963) have determined ^{187}Os/^{186}Os and ^{187}Re/^{186}Os ratios today and, by extrapolation of the regression line to a value of ^{187}Re/^{186}Os $= 0$, have obtained a primordial ^{187}Os/^{186}Os ratio of 0·831, the slope of the line giving an age of $4·0 \times 10^9$ years corresponding to a half-life of $4·3 \times 10^{10}$ years.

The rhenium and osmium contents of stony meteorites, 60×10^{-9} g/g and 710×10^{-9} g/g respectively, determined by neutron activation analysis have been described by Morgan and Lovering (1964).

Apart from meteorite studies the Os–Re method, when applied to terrestrial material, can date only the age of molybdenum mineralization. In terms of ore deposits, this may not give the age of general mineralization. Apart from mineral deposits and a few rare minerals occurring in pegmatites, the method is restricted to particular geological systems and cannot be used over the range covered by the Rb–Sr method. Ore minerals represent the final product of a complicated geochemical system, and at present the Re–Os method can be used only to investigate this system as there is practically no data available for the Re, Os concentrations and ^{187}Os/^{186}Os ratios in igneous, metamorphic and sedimentary rocks. It would appear that the rare earths and Nb–Ta minerals found in most alkaline rocks may provide suitable material for dating by this method. Results obtained by Hirt *et al.* (1963) for terrestrial samples are given in Table 32.

TABLE 32. Molybdenite ages

Sample	Re content (ppm)	^{187}Os radiogenic $\times$ 100 ^{187}Re	Age (million years)	Accepted age (million years)
Chartreris I	3·04±0·39	2·30±0·32	1050±50	1000–1100
Chartreris II	6·42±1·0	1·47±0·26	1050±50	
Ottawa, Canada				
Römteland, Norway	15·0±0·7	1·72±0·14	920±45	920
Ylitornio II, Finland	44·0±3·8	3·09±0·28	1780±100⎫	
Ylitornio I, Finland	30·6±2·7	3·06±0·28	1780±100⎭	1780±100
Galway, Ireland	15·2±0·9	0·61±0·04	365±11	365±11
Mätäsvaara	17·7±1·6	4·45±0·41	2700±130	2700 (zircon)
Finland				1800 (biotite)
Mapoka River,	33·5±0·6	4·74±0·11	3000±150	3000
Sierra Leone				
Herefoss, Norway	60·6±4·8	1·36±0·11	1000±100	1000
Swaziland, S. Africa	3·41±0·56	2·47±0·42	1940±160	1940

From Hirt *et al.* (1963).

Some of the very interesting problems presented by the Re–Os method are as follows.

(a) Establishing the rhenium and osmium content of the common rock-forming minerals with emphasis upon the presence of rhenium rich phases in the sulphide minerals. For the latter the electron micro-probe method would appear to be admirable.

(b) The separation of sufficient quantities of osmium from sulphides and common silicate minerals for mass spectrometric analysis. Volatilization techniques similar to those used in trace lead isotope analysis (see p. 172) may be an answer to this problem. With larger amounts of osmium, it may be possible to use a solid compound for solid-source mass spectrometry, although the high ionization potential for osmium may not make this possible. Failing this, it may be possible to obtain isotope ratios by means of radionuclides produced by cyclotron irradiation, similar to those described by Cobb (1964) for lead.

(c) A comparison of the isotopic composition of osmium and lead separated from the same molybdenite sample.

CHAPTER 7

URANIUM–THORIUM–LEAD METHOD

I. Uranium and Thorium Method of Age Determination

The radioactive decay of uranium and thorium to stable isotopes of lead, shown in Table 33, has often been regarded as standard method with which other methods may be compared. However, the apparent direct parent–daughter relation is often upset by the loss or addition of either uranium, thorium, lead, or one of the intermediate members of the radioactive series. An added problem is the presence of common lead in radioactive minerals, although in many cases this can be estimated by measuring the isotopic composition of cogenetic galenas. Large errors can be introduced in this estimate for young samples or when a galena has been recrystallized at a later period.

Apart from minerals containing major amounts of uranium and thorium, accessory minerals commonly found in granites, such as monazite, xenotime, orthite, pyrochlore and zircon, have proved suitable for dating.

Uranium consists of three isotopes, ^{238}U, ^{235}U and ^{234}U, the natural abundances of which are $99 \cdot 2739 \pm 0 \cdot 0007\%$, $0 \cdot 7204 \pm 0 \cdot 0007\%$ and $0 \cdot 0057 \pm 0 \cdot 0002\%$ (Lounsbury, 1956), respectively. Thorium is monoisotopic, consisting of ^{232}Th.

TABLE 33(a). The ^{238}U (uranium) series

Radioelement	Nuclide	Manner of decay	Half-life
Uranium I	^{238}U	α	$4 \cdot 51 \times 10^9$ years
Uranium X 1	^{234}Th	β^-	24·10 days
Uranium X 2	^{234m}Pa	β^-	1·175 min
Uranium Z	^{234}Pa	β^-	6·66 h
Uranium II	^{234}U	α	$2 \cdot 48 \times 10^5$ years
Ionium	^{230}Th	α	$8 \cdot 0 \times 10^4$ years
Radium	^{226}Ra	α	1622 years
Radium emanation, radon, niton	^{222}Rn	α	3·8229 days
Radium A	^{218}Po	α, β^-	3·05 min

99·98% α | 0·02% β^-

| Radium B | ^{214}Pb | β^- | 26·8 min |
| Astatine | ^{218}At | α, β^- | 1·5–2 sec |

99·9% α | 0·1% β^-

| Radium C | ^{214}Bi | β^-, α | 19·7 min |

0·04% α | 99·96% β^-

Radon	^{218}Rn	α	0·019 sec
Radium C'	^{214}Po	α	$1 \cdot 64 \times 10^{-4}$ sec
Radium C''	^{210}Tl	β^-	1·32 min
Radium D	^{210}Pb	β^-	19·4 years
Radium E	^{210}Bi	β^-, α	5·013 days

5×10^{-5}% α | 99+ % β^-

Thallium	^{206}Tl	β^-	4·19 min
Radium F	^{210}Po	α	138·401 days
Radium G	^{206}Pb	stable	—

Decay constant of ^{238}U is $1 \cdot 54 \times 10^{-10}$ year^{-1} (Fleming, Ghioroso and Cunningham, 1952).

TABLE 33(b). The ^{235}U (actinium) series

Radioelement	Nuclide	Manner of decay	Half-life
Actinouranium	^{235}U	α	$7 \cdot 1 \times 10^8$ years
Uranium Y	^{231}Th	β^-	25·64 h
Protactinium	^{231}Pa	α	$3 \cdot 43 \times 10^4$ years
Actinium	^{227}Ac	β^-, α	21·6 years
Actinium K	^{223}Fr	β^-, α	22 min
Radioactinium	^{227}Th	α	18·17 days
Astatine	^{219}At	α, β^-	0·9 min
Actinium X	^{223}Ra	α	11·68 days
Bismuth	^{215}Bi	β^-	8 min
Actinon	^{219}Rn	α	3·92 sec
Actinium A	^{215}Po	α, β^-	$1 \cdot 83 \times 10^{-3}$ sec
Actinium B	^{211}Pb	β^-	36·1 min
Astatine	^{215}At	α	$\sim 10^{-4}$ sec
Actinium C	^{211}Bi	α, β^-	2·16 min
Actinium C″	^{207}Tl	β^-	4·79 min
Actinium C′	^{211}Po	α	0·52 sec
Actinium D	^{207}Pb	stable	—

Branching fractions shown in the decay scheme: Actinium → 1·2% (α) to Actinium K; 98·8% (β^-) to Radioactinium. Actinium K → $\sim 6 \times 10^{-3}\%$ (α) to Astatine; 99+% (β^-) to Actinium X. Astatine → $\sim 97\%$ (α) to Bismuth; $\sim 3\%$ (β^-) to Actinon. Actinium A → 99+% (α) to Actinium B; $5 \times 10^{-4}\%$ (β^-) to Astatine. Actinium C → 99·7% (α) to Actinium C″; 0·3% (β^-) to Actinium C′.

Decay constant of ^{235}U is $9 \cdot 71 \times 10^{-10}$ year^{-1} (Fleming, Ghioroso and Cunningham, 1952).

The manner of decay and half-lives for uranium and thorium are given in Table 33. Half-life determinations are made by measuring the specific α-particle radioactivity. The half-life for ^{238}U is known with some certainty, although that for ^{235}U is more difficult to measure because of α-particle interference from ^{238}U and ^{234}U; the uncertainty in this decay constant is probably about 2%.

TABLE 33(c). The ^{232}Th (thorium) series

Radioelement	Nuclide	Manner of decay	Half-life
Thorium ↓	^{232}Th	α	$1 \cdot 39 \times 10^{10}$ years
Mesothorium I ↓	^{228}Ra	β⁻	6·7 years
Mesothorium II ↓	^{228}Ac	β⁻	6·13 h
Radiothorium ↓	^{228}Th	α	1·910 years
Thorium X ↓	^{224}Ra	α	3·64 days
Thorium emanation, thoron ↓	^{221}Rn	α	51·5 sec
Thorium A ↓	^{216}Po	α	0·158 sec
Thorium B ↓	^{212}Pb	β⁻	10·64 h
Thorium C	^{212}Bi	β⁻, α	60·5 min
36·2% | 63·8% α↓ Thorium C″	^{208}Tl	β⁻	3·10 min
β⁻ Thorium C′	^{212}Po	α	$3 \cdot 04 \times 10^{-7}$ sec
Thorium D	^{208}Pb	stable	—

Decay constant of ^{232}Th is $4 \cdot 99 \times 10^{-10}$ year^{-1} (Picciotto and Wilgain, 1956). Table 33(a), (b), (c) taken from Rankama (1963).

During the crystallization of magmas, uranium, thorium and lead show dissimilar geochemical behaviour, the most outstanding difference being the chalcophile character of lead. However, in dealing with radioactive minerals, conventional geochemical characteristics do not apply, as the lead is present by nature of the radioactive decay of uranium and thorium, and not through natural processes of chemical differentiation.

II. Some Selected Analytical Procedures for Uranium, Thorium and Lead

The need for improved techniques for the determination of uranium, thorium and lead associated with the growth of atomic energy research has resulted in the publication of a vast amount of literature describing these methods. It is beyond the scope of this book to describe and discuss the merits of all these methods; instead, a general review of methods relevant to isotope and geochemical studies will be given.

The analysis of these elements in geological material presents many problems, on account of the diverse chemical composition of rocks and minerals. Modern separation techniques have greatly eased the task of the analyst, but extreme precautions are necessary when determining these elements at microgram levels. Loss in chemical yields, contamination introduced by these elements in chemical reagents, glassware and from the atmosphere have necessitated special techniques.

The choice of analytical method is governed by the concentration of the uranium, thorium and lead present, which may range from a fraction of a part per million to almost a pure form. In chemical methods, it is first necessary to separate the element free from any others and, provided that this separation is achieved, there is a variety of methods for the final end determination.

A. CHEMICAL SEPARATION

The common rock-forming minerals, feldspar, pyroxene, amphibole and mica, are dissolved in mixed acids, but in many rocks much of the uranium, thorium and lead is contained within such refractory accessory minerals as zircon, and in rare-earth, niobium and tantalum minerals. The complete dissolution of these minerals may be difficult, and it is necessary to use fused alkalies (Na_2CO_3, NaOH, Na_2O_2 or $K_2S_2O_7$) to accomplish this. The detailed chemistry of U, Th, Pb has been given by, Katz and Rabinowitch (1951) Hillebrand, Lundell, Bright and Hoffman (1953), Wilson and Wilson (1959), and Grimaldi, May, Fletcher and Titcomb (1954).

Foremost in chemical separation are the use of ion-exchange separations (Samuelson, 1953; Kraus and Nelson, 1956), partition chromatography (Kember, 1952; Williams, 1952) and liquid–liquid extraction using 2-thenoyl-trifluoroacetone (T.T.A.) (Hagemann, 1950; Posanker and Foreman, 1961).

Thorium is uniquely separated by use of mesityl oxide (Levine and Grimaldi, 1950), while dithizone extractions are used in the separation of lead. In addition, lead can be effectively separated from a rock by heating (pyrochemistry, see p. 171).

B. SPECTROPHOTOMETRY

While many reagents are capable of giving a specific colour with uranium, the majority of them are neither sensitive nor specific enough for solutions containing less than 5 ppm U (Rodden, 1950). Yoe, Will and Black (1953) have discussed various organic compounds with regard to their suitability as colourimetric reagents. Dibenzolmethane (1,3-diphenyl-1,3-propanedione) has been used with success for concentrations between 0·2 and 10 ppm, with a limit of sensitivity of 0·05 ppm. Cheng (1958) has described the use of PAN (1-(2-pyridylazo)-2-naphthol) when used in conjunction with strong complexing agents such as ethylenediamine tetra-acetic acid (EDTA) and cyanide to form a deep red finely divided precipitate in ammoniacal solutions. An advantage of this method is that, within limits, trace amounts of uranium can be determined in the presence of other elements without prior separation.

Thorium is best determined by use of Thoronol (1-(O-arsonophenylozol)-2-naphthol-3), but it is essential to remove any contaminating elements (Thomson, Perry and Byerly, 1949; Grimaldi and Fletcher, 1956; Grimaldi and Jenkins, 1957).

C. RADIOACTIVATION ANALYSIS

1. Uranium (Grindler, 1962)

Neutron-induced fission of ^{235}U gives rise to a whole suite of fission products any of which can, in theory, be used to determine the total uranium content of a sample. It is necessary to make the assumption that the isotopic abundance of ^{235}U is invariant in nature. The fission product ^{140}Ba was selected by Smales (1952, 1955) as the most suitable for radiochemical assay. It has a high fission yield of 6·1% and is not produced by any non-fissile naturally occurring nuclide. However, in geological materials, ^{140}Ba with a half-life of 12·8 days is generally contaminated by ^{131}Ba with a half-life of 12·0 days produced by the irradiation of natural ^{130}Ba in the sample. However, ^{140}Ba decays to the daughter ^{140}La with a half-life of 46 h which can be used for the positive determination of uranium in the presence of other radioactive barium nuclides. This method requires a preliminary separation of radiochemically pure barium followed by the separation of the daughter product ^{140}La.

König and Wänke (1950) have extracted fission product xenon (^{133}Xe, $t_{\frac{1}{2}}$ 5·3 days; ^{135}Xe, $t_{\frac{1}{2}}$ 9·2 h) after neutron irradiation, using a glass vacuum extraction system. The use of fission products is less sensitive than that based on the ^{238}U (nγ) reaction, because of low natural abundance and fission yields of ^{235}U. The reaction ^{238}U (nγ) ^{239}U, $t_{\frac{1}{2}}$ 23 min has been used by Decat, Zanten and Leliaert (1963) for the determination of uranium. A limit of detection of 5×10^{-11} g U can be obtained by counting the 74 keV γ-radiation, but the short half-life necessitates rapid chemical separations. This can be overcome by using the reaction ^{238}U$\rightarrow$^{239}U$\rightarrow$^{239}Np, $t_{\frac{1}{2}}$ 2·3 days. The short half-life of 2·3 days of the daughter product ^{239}Np is sufficient to survive the chemical separations necessary to obtain it in a radiochemically pure form. The limit of detection by this method is 10^{-10}–10^{-13} g U. Loss of neptunium during the chemical procedures is controlled by using the α-emitting ^{239}Np for the determination of chemical yields (Morgan and Lovering, 1963).

The heterogeneous distribution of uranium in rocks and minerals often requires a large number of analyses, which are not easy to obtain by conventional activation methods. This problem has perhaps been overcome by counting the delayed neutrons emitted from the uranium atom after thermal neutron irradiation. This method was first described by Ecko and Turk (1957) and later by Hoffman (1961) and by Dyer, Emery and Leddicotte (1962).

The delayed neutrons are emitted from nuclei which have been left in a highly excited state by the negatron decay of fission-produced parent nuclei. The neutrons are emitted instantaneously after the negatron is emitted, and the term precursor is given to the radionuclide which decays by negatron emission to produce a neutron-emitting radionuclide. There are at least seven precursors with half-lives ranging between 1·6 and 54·5 sec. In this technique, a sample is loaded into a fast delivery and return system (rabbit) by which it is possible to project by air pressure the sample from a laboratory into the reactor, to leave it there for an accurately determined period and then to return it to the laboratory. On its return, the sample is ejected into a neutron counter and the accumulated counts taken over a period of about 10 min are recorded. The same procedure is repeated for standards and samples, each complete analysis taking about 15 min. With short irradiation in a neutron flux of 6×10^{13} n cm^{-2} sec^{-1}, and using a BF$_3$ neutron counter with an efficiency of about 5%, the limit of measurement for natural uranium is about 7×10^{-7} g (Dyer et $al.$, 1962). The method is almost free from any interference from other radionuclides, is extremely rapid and has a precision of about $\pm 3 \cdot 0\%$ expressed as the relative standard deviation. Apart from the determination of uranium, the use of fast neutrons can be used to measure the thorium content of geological material in the presence of uranium.

2. Thorium (Hyde, 1960)

Thorium can be determined by the measurement of ^{233}U or ^{233}Pa radionuclides, the decay schemes of which are:

$$^{232}\text{Th} \rightarrow {}^{233}\text{U} \rightarrow {}^{233}\text{Pa}$$
$$\begin{array}{cc} 22\cdot4 & 27\cdot0 \\ \text{min} & \text{days} \end{array}$$

By performing rapid chromatographic separations, Jenkins (1955) used ^{233}U, but the short half-life is not suitable when protracted chemical separations are necessary to achieve radiochemical purity. In this instance, use of the daughter product ^{233}Pa is to be preferred (Morgan and Lovering, 1963).

3. Lead (Gibson, 1961)

Lead can be determined by the reactions ^{208}Pb (nγ) ^{209}Pb and ^{204}Pb (n2n) ^{203}P but a low cross-section for these reactions places a practical limit of sensitivity of about 0·01 ppm when using a neutron flux of 10^{13} n cm^{-2} sec^{-1} (Hama-guchi, Reed and Turkevich, 1957; Reed, Kigoshi and Turkevich, 1960; Tilton and Reed, 1960). In addition, it is necessary to know the isotopic composition of lead, which is variable in different minerals, depending upon the previous geochemical association with uranium and thorium. However, in spite of these difficultues, the method is free from contamination that can affect more refined procedures such as that of isotope dilution analysis.

Cobb (1964) has determined the isotopic abundances of lead in meteorites by bombardment with 30-MeV α-particles to induce an (α2n) reaction of the lead isotopes, as shown in Table 34. The activity of ^{210}Po and ^{208}Po are

TABLE 34. Products from (α2n) reaction on lead

Lead isotope	Polonium isotope	Half-life	Decay Characteristic
204	206	8·8 days	5%α, 95%e capture
206	208	2·93 years	α 5·11 MeV
207	209	~200 years	α 4·89 MeV
208	210	138 days	α 5·30 MeV

determined by counting the α-particles in a gridded ion chamber. By this method, it is possible to measure the relative abundances of ^{204}Pb, ^{206}Pb and ^{208}Pb down to a total lead concentration of 0·1 ppm, and the ^{208}Pb/^{206}Pb ratio at a level of 0·01 ppm. Actual concentrations cannot easily be measured because of the absence of a flux monitor and the lack of a suitable tracer to correct for chemical yields.

D. Determination of Uranium and Thorium by means of Natural Radioactivity

A direct measurement of the uranium and thorium content of a rock, without any prior chemical separations, may be made by γ-spectrometry. The natural γ-activity emitted from uranium and thorium is hardly absorbed when passing through natural silicates. It is only necessary to powder a rock sample and compact it about a γ-sensitive scintillation crystal. The sensitivity of detection may be improved by increasing the sample size, using a large (3 inch × 3 inch) scintillation crystal and increasing counting times. The precision of the method depends upon the composition of the sample and the equipment used but, in general, a precision of 10–15% can be obtained.

Thorium is determined from the γ-emission of 2·62 MeV peak ([208]Tl), uranium from that of 1·76 MeV ([214]Bi); descriptions of various techniques used have been described by Eicholz, Hilborn and McMahon (1953), Hurley (1956), Adams, Richardson and Templeton (1958), Whitefield, Rogers and Adams (1959) and Bloxham (1962). It is also possible to measure the potassium content of rocks by counting the 1·47 MeV peak of naturally radioactive [40]K.

On the assumption that in a sample the uranium/thorium ratio is a known constant, and that radioactive equilibrium exists, the uranium or thorium content can be determined, even at low levels, by counting the total α-particle activity. In other cases, when the U/Th ratio is unknown, the same technique can give meaningful results in terms of the equivalent uranium or thorium content. A refinement of this technique is the application of α-spectroscopy which has been applied to a study of the uranium content of minerals by Facchini, Forte, Malvicini and Rossini (1956).

E. FLUORIMETRIC ANALYSIS

(Grimaldi, 1952; Kinser, 1954; Adams and Maeck, 1954)

With appropriate techniques, the fluorimetric method of analysis for uranium has a sensitivity comparable to that of neutron activation analysis ($10^{-4}\,\mu g$). After dissolution of the sample, uranium is extracted into an appropriate organic phase (i.e. ethyl acetate), back extracted into water and, after evaporation to a small volume, fused with a mixed carbonate–fluoride flux to form a thin pad. The pad is then subjected to ultra-violet radiation produced by a high-intensity mercury lamp. The ultra-violet light is passed through a filter to pass 3650 Å quanta, which is considered the most desirable wavelength for the production of uranium fluorescence. The fluorescence produced by the sample passes through a secondary filter with a low wavelength cut-off at ~ 5545 Å, which is the emission peak of uranium fluorescence Following this, the emission is then registered by means of a phototube or photomultiplier and the uranium concentration read by the deflection produced on a galvanometer.

The sample uranium fluorescence can be enhanced or, more commonly, quenched by the presence of contaminating elements. This can be corrected for by fusing a known amount of uranium of known fluorescence in the sample pad such that the fluorescence contribution of the sample is completely masked. The fluorescence is then re-measured to determine the amount of quenching shown by the added uranium. The loss of uranium during the chemical procedures can be corrected by use of [233]U as a radioactive tracer.

An alternative approach is one of infinite dilution, by which the sample is diluted such that the effect of quenching elements is reduced to an insignificant amount, but the fluorescence of uranium can still be measured.

The fluorimetric method of analysis is best suited to problems requiring a large number of analyses, and when accuracy is not so vital. This can be of particular use in the selection of suitable material prior to detailed radiometric dating.

Thorium has also been determined at sub-microgram quantities (Sill and Willis, 1962) by use of morin in an alkaline solution of EDTA.

F. AUTORADIOGRAPHIC TECHNIQUES

Autoradiographic methods can be used to obtain a "visual picture" of the distribution of α-radioactivity in rocks and minerals.

In autoradiographic techniques, a photographic film is placed in contact with a polished radioactive sample. As the α-particles penetrate the photographic emulsion, they leave a trace of developable silver grains. When the film is developed, the track of the α-particles is recorded as a thin line of silver grains the length of which corresponds to the energy of the particular α-particle in the emulsion. For highly radioactive samples, a simple photographic plate, sensitive to α-, β- and γ-radiation, will suffice to obtain the distribution of radioactivity. In more refined techniques, a liquid photosensitive emulsion, sensitive only to α-particles, is poured over a thin section of a rock, it is then dried and stored for a convenient length of time; an exposure of 3 weeks is generally suitable for granites. After developing the emulsion in contact with the thin section, the actual distribution of α-particle activity can be observed by using a conventional microscope and focusing down through the emulsion to the emulsion-thin section interface. It is possible to follow a track and observe the source of the α-particle; this may be from the body of a mineral, inclusion in a mineral, along crystal cracks or grain boundaries.

The distribution of radioactivity in the International Standard G1, is given in Table 35 and the α-activity emitted from an inclusion of apatite in feldspar from G1 is given in Fig. 33. In theory, it is possible to distinguish between uranium and thorium by the length (mean energy) of the track; but, in practice, this is difficult. Nuclear emulsion studies on a wide range of rocks have produced the following five results.

1. The level of radioactivity can be correlated, within limits, with variations in the major elements although, at very low concentrations, the uranium and thorium content appears to be independent of mineral composition.

2. Most of the radioactivity in igneous rocks is contained within accessory minerals and inclusions within the major minerals. The distribution of radioactivity in both is heterogeneous with well defined points of high activity superimposed upon a more homogeneous background.

3. In some samples the radioactivity is found along cracks in mineral and along intercrystal boundaries. In many cases this can be identified with films of iron oxides.

4. Emanation of radon and thoron is higher from altered minerals, or minerals that have been affected by solid state phase changes.

5. Zircons are often surrounded by narrow zones that are significantly radioactive.

FIG. 33. α-Particle activity from an apatite inclusion in feldspar of standard G1. The α-particle radioactivity is registered in a nuclear emulsion covering a thin section of the rock. The tracks are approximately 15μ in length. (Photograph, E. I. Hamilton.)

In uranium–thorium dating, autoradiographic studies can be used for selecting suitable minerals. They would also appear to have a potential use in tracing the development of galenas with anomalous isotopic compositions.

α-Particle autoradiography applied to a study of natural silicates has been described by Yagoda (1949), Picciotto (1950), Bowie (1954) and Hamilton (1959a, b 1960, 1964). β-Particle autoradiography is also possible using β-sensitive photographic emulsions. An interesting extension of this work is a study of induced β-activity in rocks and minerals after neutron irradiation; an example where this method would be useful is a study of the distribution of silver in galenas. Silver is often present in moderate concentrations and would be highly radioactive compared to lead, the cross-section for activation of

TABLE 35. The distribution of radioactivity and uranium in three thin sections of International Granite Standard G1

Mineral	Total α-particle activity			% Activity from rock			% Activity from mineral			α-Particle activity ×10⁻⁵ (α cm⁻² sec⁻¹)			Average uranium content[1] ×10⁻⁶ g per g
	1	2	3	1	2	3	1	2	3	1	2	3	
FELDSPAR	410	249	309	30	19·24	11·95	100·0	100·0	100·0	9·73	5·54	6·48	0·27
Feldspar	68	2	31	5·0	0·16	1·19	16·6	1·0	10·0	1·60	0·06	0·65	0·03
Inclusions	280	242	250	20·5	18·69	9·68	68·0	97·0	81·0				
Cracks	62	5	28	4·5	0·39	1·08	15·4	2·0	9·0				
QUARTZ	94	68	48	6·9	5·25	1·86	100·0	100·0	100·0	4·83	3·76	2·89	0·16
Quartz	29	2	10	2·1	0·16	0·39	30·0	3·0	21·0	1·45	0·11	0·11	0·03
Inclusions	50	59	26	3·7	4·55	1·01	54·0	86·7	54·0				
Cracks	15	7	12	1·1	0·54	0·46	16·0	10·3	25·0				
Biotite[2]	219	77	197	16·0	5·9	7·62	100·0	100·0	100·0	73·8	39·0	98·2	0·28
Muscovite	4	8	27	0·4	0·62	1·01	100·0	100·0	100·0	5·93	11·87	43·0	
Apatite	4	0	10	0·4	0	0·39	100·0	100·0	100·0				
Zircon	61	6	4	4·5	0·47	0·15	100·0	100·0	100·0				
Accessory A+ore	50	0	37	3·7	0	1·92	100·0	100·0	100·0				
Intercrystal	74	67	69	5·4	5·18	2·67	100·0	100·0	100·0				
Accessory inter-crystal	34	259	0	2·5	20·10	0	100·0	100·0	100·0				
Accessory B	220	117	59	16·0	9·05	2·28	100·0	100·0	100·0				
Accessory C	194	443	1789	14·2	34·24	69·18	100·0	100·0	100·0				
Total area (cm²)	2·6	2·6	2·2										

which is low. Other elements worth further study are Mn, rare earths (Sm, Eu, Gd), Sc, Ti, Cr, Co and Cu.

G. Mass Spectrometry

Uranium may be determined by surface ionization techniques using uranium in the nitrate form, or in gas analysis by means of uranium hexafluoride. Porous single tantalum filaments have been used, but triple filaments have proved to be up to 200 times more efficient in ion production (Inghram and Chupka, 1953; Lounsbury, 1956).

The use of solid samples requires only microgram amounts of uranium compared with very much greater amounts when uranium hexafluoride gas is used. In gas analysis, memory effects can be reduced using a suitable ion source and inlet system, but errors are introduced by the presence of fluorocarbon compounds and, among others, by the deposition of solid uranium with its high radioactivity. The isotopic analysis of uranium hexafluoride with high precision using improved source design has been described by Bishop, Davidson, Evans, Hamer, McKnight and Robbins (1960). For isotope dilution analysis a tracer is used consisting of enriched ^{235}U, and the natural isotopic composition is assumed to be invariant, viz. $99 \cdot 27\%$ ^{238}U. The chemical and physical procedures used in age studies have been described by Tilton (1951) and Cobb and Kulp (1961).

Thorium consists of only one isotope, ^{232}Th and, consequently, the determination of thorium by isotope dilution is only possible if sufficient amounts of an artificially produced isotope are available. ^{230}Th has been used as a tracer, but severe precautions are needed because of its toxicity.

H. Other Methods

The determination of uranium, thorium and lead by optical spectroscopy has been described by Waring and Mela (1953), Waring and Worthing (1953), Vinogradov (1956) and Ahrens and Taylor (1963).

X-Ray fluorescence analysis has been applied to geochemical studies only in recent years and has shown much promise. With appropriate matrix corrections and choice of standards, both high accuracy and precision can be obtained in the analysis of total uranium, thorium and lead in rocks and minerals over a wide range of concentration.

III. Uranium–Thorium Age Method

The principles of this method were founded and developed prior to mass-spectrometric analysis. This early, but very important, work involved the gravimetric determination of total uranium, thorium and lead, while the atomic weight of the separated lead was estimated by direct weighing (Knopf et al., 1931). Although these chemical separation methods are still used, they have been supplanted by mass-spectrometric measurements. The general age equations and their derivation have been derived by Knopf et al. (1931) and Keevil (1939), while Greenhalgh and Jefferey (1959) and Stieff and Stern (1961) have published tables for the solution of these age equations.

Uranium and thorium lead ages can be calculated directly from the general law of radioactive decay; uranium and thorium constitute the parent atoms and lead the daughter atoms. The growth of radiogenic lead-206, -207 and -208 can be described by the following equation illustrated by the growth of lead-206:

$$\underset{\text{measured}}{{}^{206}\text{Pb}} = \underset{\text{original}}{{}^{206}\text{Pb}} + {}^{238}\text{U}(e^{\lambda t} - 1) \tag{18}$$

dividing through by non-radiogenic ${}^{204}\text{Pb}$ we have

$$\underset{\text{measured}}{\frac{{}^{206}\text{Pb}}{{}^{204}\text{Pb}}} = \underset{\text{original}}{\frac{{}^{206}\text{Pb}}{{}^{204}\text{Pb}}} + \frac{{}^{238}\text{U}}{{}^{204}\text{Pb}}(e^{\lambda t} - 1) \tag{19}$$

Equation (19) is one of a straight line with ordinate ${}^{206}\text{Pb}/{}^{204}\text{Pb}$, abscissa ${}^{238}\text{U}/{}^{204}\text{Pb}$, and slope $(e^{\lambda t} - 1)$.

Ages can be calculated from the following equations.

$$\frac{{}^{206}\text{Pb}}{{}^{238}\text{U}} = e^{\lambda 238 t}$$

$$\frac{{}^{207}\text{Pb}}{{}^{235}\text{U}} = e^{\lambda 235 t}$$

$$\frac{{}^{207}\text{Pb}}{{}^{206}\text{Pb}} = \frac{e^{\lambda 235 t}}{k\, e^{\lambda 238 t}} \qquad \text{where } k = \frac{{}^{238}\text{U}}{{}^{235}\text{U}} \text{ today}$$

$$\frac{{}^{208}\text{Pb}}{{}^{232}\text{Th}} = e^{\lambda 232 t}$$

If the sample has remained a closed system throughout its history, then all three uranium ages should be identical (concordant). The early work of Nier (1939, 1941) showed that all three ages need not necessarily agree and that discordant age patterns are common. However, the patterns could provide valuable information as they might reflect the pre- or post-crystallization history of a mineral. Apart from concordant and discordant ages, it is also possible to obtain an apparently concordant age pattern which is not the real age of the mineral (Wetherill, 1956).

Tilton has suggested that if the various U–Pb ages differ by more than 10% they may be regarded as discordant. The discordant patterns for uranium minerals commonly show the sequence

$$^{207}Pb/^{206}Pb > ^{207}Pb/^{235}U > ^{206}Pb/^{238}U \gg ^{208}Pb/^{232}Th,$$

other radioactive minerals may show the sequence

$$^{207}Pb/^{206}Pb \gg ^{207}Pb/^{235}U > ^{206}Pb/^{238}U > ^{208}Pb/^{232}Th,$$

while a few minerals show the reversed sequence

$$^{206}Pb/^{238}U > ^{207}Pb/^{235}U \gg ^{207}Pb/^{206}Pb;$$

often the lead–thorium ages are the lowest found.

VI. Discordant Ages

Ahrens (1955a,b) was the first to observe a linear relationship when the $^{207}Pb/^{235}U$ ratio was plotted against the $^{206}Pb/^{238}U$ ratio for discordant age sequences. The apparent regularities in discordant age patterns were regarded as being related to a fundamental primary process which operated to a varying degree, but at a uniform rate since the mineral was formed. Wetherill (1956) enlarged upon Ahrens's mode of presentation to explain discordant ages in terms of single or multiple episodes of lead–uranium fractionation. In place of Ahrens's plot of the mole ratio of $^{207}Pb/^{235}U$ against $^{206}Pb/^{238}U$, Wetherill introduced a curve called "concordia" (see Fig. 34), which represents the locus of all points having equal $^{207}Pb/^{235}U$ and $^{206}Pb/^{238}U$ ages. Thus if both uranium isotope ratios fall on concordia, the ages will be concordant, i.e. $^{206}Pb/^{238}U$ age = $^{207}Pb/^{235}U$ age. Any point that does not lie on the curve represents a discordant age.

Both ages will be "concordant" if the following conditions are fulfilled.

1. The mineral has acted as a closed chemical system and there has been no gain or loss of uranium or lead; it is assumed that the rate of ^{206}Pb and ^{207}Pb loss or gain is identical.

2. Corrections have been made for any original radiogenic lead present in the mineral.

3. The chemical and isotope measurements have been made correctly, and the decay constants for ^{238}U and ^{235}U are known.

The graphical solution to discordant uranium–lead ages was developed by Wetherill (1956), who also described extensions of the method for calculating the effect of the presence of primary radiogenic lead and of constant loss of intermediate decay products.

All analyses used in a "concordia" plot have to be corrected for any non-radiogenic lead present in the sample. Concordant uranium ages are considered to be the correct ages, and indicate that the minerals have not been altered by any later event.

The mathematical proof of concordia plots are complex, in particular, the extension of the initial theory, and for present purposes a proof described by Stieff and Stern (1961) is given, while the mathematically inclined reader is referred to Wetherill's original paper.

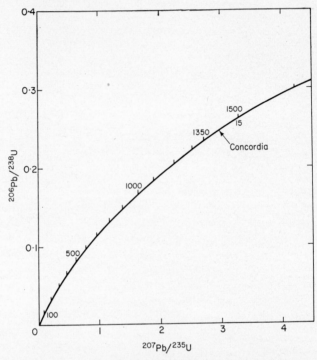

FIG. 34. Plot of $^{206}Pb/^{238}U$ against $^{207}Pb/^{235}U$, which is useful for the interpretation of dis cordant U–Pb ages. The curve is the locus for concordant ages, and the corresponding age (million years) are given. (After Aldrich and Wetherill, 1958.)

Referring to Fig. 35, assume that: (a) samples A, B and C were formed a the same time, t; (b) sample A is unaltered; (c) sample B has recently los lead or gained uranium, while sample C has recently lost uranium or gaine radiogenic lead; (d) recent loss or gain of lead or uranium will not significantl alter the $^{207}Pb/^{206}Pb$ ratios of samples B and C by isotopic fractionation c ^{206}Pb or ^{207}Pb. Then, (i) the curve concordia in terms of the mole ratio i defined by

$$N_{206}/N_{238} = e^{\lambda_{238}t}$$
$$N_{207}/N_{235} = e^{\lambda_{235}t}$$

or values of t between 0 and infinity; (ii) the point A lies on the curve concordia and all three ages are equal; (iii) the points B and C do not lie on the
concordant age curve because their $^{206}Pb/^{238}U$ and $^{207}Pb/U^{235}$ are not equal;
(iv) the slope of the line connecting point A with the origin multiplied by the

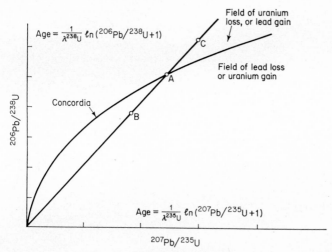

FIG. 35. Plot of $^{206}Pb/^{238}U$ against $^{207}Pb/^{235}U$ for uranium minerals which have recently
lost or gained lead or uranium. (After Stieff and Stern, 1961.)

atom ratio of the present day relative abundance of $^{235}U/^{238}U$, 1 : 137·8, must
give the $^{207}Pb/^{206}Pb$ ratio of the radiogenic lead in sample A

$$(N_{207}/N_{235}) \times (1/137·8) \times N_{238}/N_{206} = N_{207}/N_{206};$$

(v) the slope of the line passing through the origin and the points B and C
multiplied by 1/137·8 will give the N_{207}/N_{206} ratio of B and C; (vi) the radiogenic $^{207}Pb/^{206}Pb$ ratios of A, B and C are equal (assumptions (a) and (c));
therefore, the points B and C must lie on a single line passing through
the origin and intersecting the curve concordia at point A; (vii) the concordant age for samples B or C corrected for recent loss or gain of lead or
uranium will therefore be given by the N_{207}/N_{235} and N_{206}/N_{238} ratios at the
intersection of the line passing through the origin and points B or C and the
curve concordia.

The recent loss or gain of lead and uranium can also be obtained directly
by a graphical approach. In Fig. 35, the length of the chord OA is directly
proportional to the ratio of the total number of atoms of lead and uranium
required to produce a concordant age. The segment OB lying on the chord
OA, and the segment OC lying on OA extended are also proportional to the
ratios of the lead and uranium remaining after recent processes of alteration.

Therefore the length of the line segment BA to OA × 100, will give the percentage of lead lost, or uranium gained recently. The ratio of the length of the line OC to the line OA × 100 will give the percentage of uranium lost or lead gained recently.

This graphical analysis demonstrates that plots of radiogenic lead against uranium for a suite of cogenetic uranium-bearing minerals which have been recently altered will lie on a straight line passing through the origin, the intersection of this line with concordia will equal the $^{207}Pb/^{206}Pb$ age of the sample, and the extent of the later alteration may be determined graphically.

A plot of N_{207}/N_{204} against N_{206}/N_{204} has the added advantage that no correction need be made for any contaminating lead. In such graphs, the slope of the line passing through co-ordinates of a set of cogenetic minerals will give $^{207}Pb/^{206}Pb$ ages independent of any later process of alteration.

Apart from recent loss or gain of lead or uranium, similar graphical plots can be used to describe lead or uranium lost in the past. For a detailed account of the solution of discordant lead–uranium ages the reader is referred to Keevil (1939), Wetherill (1956), Kulp and Eckelman (1957), Russell and Ahrens (1957), Stieff, Stern, Oshiro and Senftle (1959) and Stieff and Stern (1961).

V. Some Explanations of Discordant Ages

Current research is directed towards investigating the causes of discordant ages. Although there are many approaches to this complex problem, for the present purpose they may be collected under two headings.

A. The Loss or Gain of Parent, Intermediate Members or Final Products of the Radioactive Series

Holmes (1948) explained the discordant pattern of $^{206}Pb/^{238}U < ^{207}Pb/^{235}U$ ages in terms of radon loss from the minerals. However, an examination of monazite from a pegmatite (Holmes, 1955) did not show radon, but thoron loss, indicating an inhomogeneous distribution of uranium and thorium within the mineral. Kulp, Bate, Broecker (1954) and Kulp (1955) suggested that radon loss may occur at elevated temperatures, while Giletti and Kulp (1955) showed that radon loss was not serious in cases of primary uranium minerals, although high losses were associated with secondary and altered minerals. Kuroda (1955) has studied the effect of fractionation between intermediate members of the radioactive series. The ratio $^{223}Ra/^{226}Ra$ in 32 different samples from 28 localities showed a variation between 0·046 (the expected ratio from the present $^{235}U/^{238}U$ ratio) to 0·130. These results show that, even if uranium and thorium are homogeneously distributed in mineral, the lead isotope ratios can be affected as a result of fractionation between intermediate members.

Stieff and Stern (1956) have suggested that natural geological processes may be the cause of discordant patterns. They have described a mixing process for lead and uranium during primary mineralization, and also possibly by the selection of weathered or altered specimens for which the geological setting is not adequately known. Tilton and Patterson (1956) favour the loss or gain of uranium or lead, although they agree that the loss of intermediate nuclides, or the presence of primary radiogenic lead, may in some instances be equally as important.

The ease with which various members of the radioactive series can be removed from minerals can be approached in the laboratory by treating them with mild acids or water and examining the leached products. Starik, Starik and Petryaev (1955) leached uraninite and found a preferential loss of ^{233}Ra, ^{224}Ra and ^{226}Ra with respect to uranium and preferential leaching of ^{233}Ra and ^{224}Ra with respect to ^{226}Ra. For allanite, Vinogradov (1956) found a preferential loss of uranium and lead. Leaching experiments by Tilton and Nicolaysen (1955) and by Kulp and Eckelmann (1957) support the view that uranium, thorium and lead are inhomogeneously distributed in monazites. In particular, thorium lead appears to be easily leached. The loss of non-radiogenic lead ^{204}Pb described by Zhirov, Zykov and Stupnikova (1957) suggests that, in some cases, leachable common lead may have been added during a hydrothermal stage and is present on exterior mineral surfaces or along cracks. Undoubtedly, leaching experiments will provide much information concerning discordant age patterns, but the use of strong leaching reagents may lead to the actual chemical dissolution of some minerals, and such results would not be comparable to processes of natural leaching.

B. Discordant Age Patterns as a Result of Diffusion Processes

Wetherill (1956) proposed a single episode of lead loss, which can be shown in a graphical form by use of a "concordia" plot. Samples that have lost lead will lie on a chord, their position on the chord determining the amount of lead lost. A plot of the parent/daughter ratios for minerals from different localities having ^{207}Pb–^{206}Pb ages of 2300–2800 million years show a definite fit to the same chord on concordia (Fig. 36). At first sight this suggests a universal loss of lead from these old minerals about 500 million years ago. While this is possible for some of the areas, it cannot be accepted for others. The applications of diffusion processes by Wasserburg and Hayden (1954), Nicholaysen (1957) and Tilton (1960) show that the continuous diffusion of lead will result in a similar pattern but, in this case, the

6

apparent age of 500 million years is not related to a single episode of lead loss, or to a specific geological event. In addition to lead diffusion, Wetherill (1963) showed that uranium diffusion will also lead to a straight line pattern (chord to concordia), but one exhibiting a different slope to that caused by the diffusion of lead.

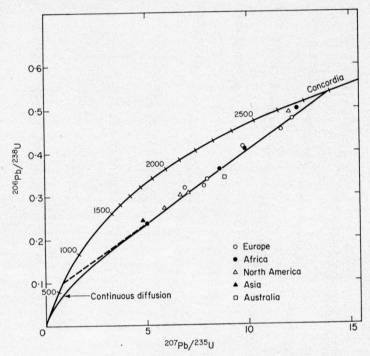

FIG. 36. Parent/daughter ratios for minerals having ^{207}Pb/^{206}Pb ages of 2300–2800 million years. The curve represents calculated values for lead loss by continuous diffusion. (After Tilton, 1960.)

It is not possible to distinguish between the loss of lead or gain of uranium; the U–Pb age may be high or low depending upon whether uranium or lead is lost. The combined effect of both lead and uranium diffusion related to a pulse of metamorphism is found to be far too complex to exhibit a regular pattern. A further mathematical treatment of diffusion in the lead–uranium system has been given by Wasserburg (1963).

C. SIGNIFICANCE OF ISOTOPIC AGES

(i) ^{206}Pb/^{238}U age. In the case of fresh uranium-rich minerals, this should approximate to the correct age. The high abundance of parent and daughter reduces any analytical error. The age is not sensitive to radon loss, but is to loss of lead or uranium.

(ii) $^{207}Pb/^{235}U$ age. The low abundance of ^{235}U presents analytical problems and, in addition, large errors can be related to uranium loss or the incorrect choice for the isotopic constitution of common lead.

(iii) $^{207}Pb/^{206}Pb$ age. For young and recent samples, the slope of the growth is very slight and any analytical error in measuring the ratios results in very large errors in the age.

The uranium–thorium–lead method of dating has been extensively applied to the dating of radioactive minerals containing appreciable amounts of uranium and thorium. Some examples of concordant and discordant ages for these minerals are given in Table 36. The choice of such samples restricts the investigator to dating specific periods of mineralization or to late stage minerals present in pegmatites. This restriction has been overcome by the application of recent analytical methods, capable of measuring very small amounts of uranium, thorium and lead in more common radioactive accessory minerals such as zircon. Tilton, Patterson, Brown, Inghram, Hayden, Hess and Larsen (1955) were the first to investigate "rock zircon" separated from a Pre-Cambrian granite, from Haliburton County, Ontario. No common lead was found, and both U–Pb ages were concordant, while the Th–Pb age was lower. A comparison of the observed U, Th and Pb content to that which would be expected showed that the granite acted as a closed system with respect to uranium and its decay products, but had been an open system with respect to thorium and its decay products.

Silver and Deutsch (1963) have made a detailed study of zircons in terms of cogenetic uranium–lead systems. Zircon, separated from the Johnny Lyon granodiorite, Arizona, was found to contain crystals of uranothorite which could be removed by acid washing; this illustrates the need for careful mineral separation. The distribution of radioactivity was determined by nuclear emulsion methods (see p. 134) and chemical stripping techniques made it possible to analyse the uranium, thorium, lead and isotopic constitution of lead from various layers of the crystals.

Discordant zircon and uraninite ages from a Pre-Cambrian granophyre suite, central Arizona, U.S.A. (Silver, 1960, 1963) yielded ages of 50–1100 million years. A "concordia" diagram, Fig. 37(a), shows a well defined pattern with an original age of 1150 ± 30 million years and a later event, supported by field studies, at 80 ± 30 million years. Prior to these studies, the age of the intrusion could be given only as pre-Devonian.

The use of zircon age studies combined with rubidium–strontium dating, in the complex Pre-Cambrian rocks of Finland, has been described by Hart, Aldrich, Davis, Tilton, Baadsgaard, Kouvo and Steiger (1963). In a study of uranium–lead systems in zircons from Pre-Cambrian rocks from Cochise Co.,

TABLE 36. Concordant and discordant uranium–thorium–lead isotope ages

Mineral	Locality	Lead isotope ages (million years)				Reference
		$^{238}U–^{206}Pb$	$^{235}U–^{207}Pb$	$^{207}Pb–^{206}Pb$	$^{232}Th–^{208}Pb$	
Concordant ages						
Zircon	Witchita Mts. Oklahoma, U.S.A.	520±12	527±10	550±30	506±12	Tilton *et al.* (1957)
Zircon	Ceylon	540±12	544±16	555±30	538±25	Tilton *et al.* (1957)
Pitchblende	Katanga, Belgian Congo	575±5	595±5	630±40	—	Eckelmann and Kulp (1956)
Uraninite	Wilberforce, Canada	1000	1015	1030	1010	Nier (1941)
Uraninite	Romteland, Norway	890	892	920	900	Kulp and Eckelmann (1957)
Samarskite	Spruce Pine, N. Carolina, U.S.A.	314	316	342	302	Eckelmann and Kulp (1957)
Thucolite	Witwatersrand, S. Africa	2110	2080	2070	—	Louw (1955)
Pitchblende	Katanga, Belgian Congo	610	615	650	—	
Discordant ages						
Zircon	Capetown, S. Africa	330±10	356±15	530±50	238±20	Tilton *et al.* (1957)
Zircon	Beartooth Mts. Montana, U.S.A.	770±25	1400±40	2580±50	—	Catanzaro and Kulp (1964)
Zircon	Montana, U.S.A.	1660±50	2380±70	3080±50	870	Catanzaro and Kulp (1964)
Zircon	Quartz Creek, Colorado	930	1130	1540	515	Carnegie Rept. 1954–55
Pitchblende	Sunshine Mine, Colorado	805±10	860±20	1035±35	—	Eckelmann and Kulp (1957)
Monazite	Huron Claim, Manitoba, U.S.A.	3220	2840	2590	1830	Nier (1939)
Xenotime	Uncompahgre, Colorado	3180	2065	1640	1100	Tilton (1956)
Euxenite	Wakefield, Quebec	620	710	1000	550	Robinson *et al.* (1963)

Arizona (Silver, 1963) has shown that a "concordia" plot can be used to resolve ages separated by about 60 million years. Zircon from a post-tectonic

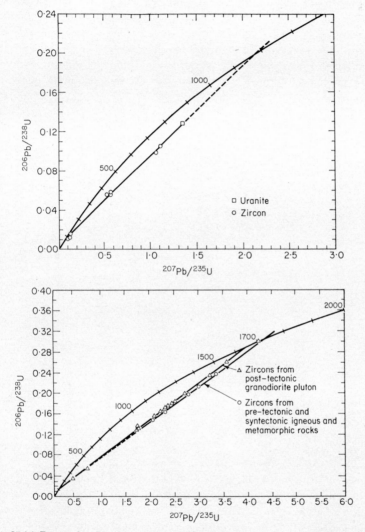

FIG. 37.(a) Parent–daughter ratios for a series of cogenetic zircons and one uraninite from a diabase–granophyre suite, central Arizona, U.S.A. (b) Zircon–uranium–lead system for Pre-cambrian rocks of the Dragoon Quadrangle, Cochise Co., Arizona. (From Silver, 1963.)

pluton and those from pre-tectonic and syntectonic igneous and metamorphic rocks were found to lie along slightly different chords, as illustrated in Fig. 37(b). When other dating methods are used, the old Pre-Cambrian ages are not seen, and have presumably been obliterated by the younger ("Laramide") event.

Catanzaro and Kulp (1964) have examined discordant age patterns for zircon in metamorphic environments. They have shown that age determinations of suites of zircons from restricted areas can provide sufficient data for geological interpretation. In some cases, radiogenic lead is apparently lost by a chemical process at low temperature. During a metamorphism, rocks containing previously formed zircons can lose all or substantial amounts of radiogenic lead.

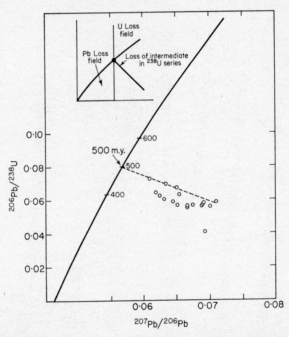

FIG. 38. Parent–daughter ratios for Swedish Kolm samples. (After Cobb, 1961.)

The effect of recycled radiogenic lead in the Blind River uranium deposit Ontario, Canada, has been discussed by Mair, Maynes, Patchett and Russell (1960). The uranium ore occurs together with quartz conglomerates at the base of a sedimentary series overlying granite and greenstone basement rock Detrital monazites and zircons gave Pb–U ages of 2500 million years, while the uraninite ores gave a lead ratio age of 1700 million years. Ages by other isotope methods suggest sediments enclosing the ore are 1700 million years or older. Lead extracted from associated pyrite, pyrrhotite, sericite and feldspar has anomalous isotopic lead ratios as a result of the addition of radiogenic lead from the uraninites.

The origin of gold and uranium in the conglomerates of the Witwatersrand system is a much disputed problem. Of the two main schools of thought, one supports a placer hypothesis, the other a hydrothermal theory of origin (Davidson, 1953). Apparent radiogenic lead age patterns from uraninites by Louw (1954, 1955) give an age of between 1850–1950 million years which would be in keeping with a detrital origin. Some reworking has occurred, as shown by the admixture of radiogenic lead in galena.

Cobb (1961) has used U–Pb dating to investigate the age of sediments. At present the method is restricted to uranium-rich sediments. It is necessary that the samples contain sufficient uranium and lead for an accurate analysis; common lead should be low, later alteration should not be present, and the uranium should accumulate during sedimentation and diagenesis. The uranium-rich Swedish Kolm shales contained excess ^{206}Pb; the discordant age pattern is attributed to the loss of radon and lead through the leaching effect of ground water. The effect of the resulting age pattern illustrated in Fig. 38 suggests a minimum age of 500 million years, obtained by the intersect of a line through the various points with concordia.

VI. Lead-210 (RaD) Dating

The lead-210 method involves measuring the ratio of one of the intermediate members of the uranium radioactive series to the daughter lead end-product. The determination of total lead and uranium is not necessary, the method of analysis is simple and rapid and, in the case of uranium minerals, a mass spectrometric analysis of the lead is not necessary, and is assumed to consist solely of radiogenic lead. The method was first proposed by Houtermans, and was applied to dating pitchblendes from the Belgian Congo by Begemann, Buttlar, Houtermans, Isaac and Picciotto (1953). Kulp, Broecker and Eckelmann (1953) determined the amount of ^{210}Pb activity or by the radioactivity of the daughter product ^{210}Bi which decays to ^{210}Po. The ^{210}Pb measurement is subject to errors by nature of its weak β-activity (^{210}Pb 0·025 MeV, $t_{\frac{1}{2}}$ 22·2 years) which necessitates counting the lead in a gaseous form (tetramethyl). The counting efficiency and the total amount of lead in the counter must be known, but this is subject to contamination by the deposition of solid members of the intermediate decay products. Bi-210 has a strong β-particle energy (Ra, E 1·17 MeV, $t_{\frac{1}{2}}$ 5·02 days), is the most suitable, and the chemical separation is very simple.

The method used for separating lead varies, depending upon the sample; generally, the lead is separated by conventional methods but, in the case of water samples, inactive lead is added as a carrier. If ^{210}Pb is used, it is necessary only to separate the total lead and, after the other radioactive isotopes have decayed, the radioactivity is measured directly on the lead salt (Kulp, Volchok and Holland, 1952b).

Lead-210 has a half-life of $21 \cdot 4 \pm 0 \cdot 5$ years, and this long half-life reduces the possibility of contamination during the chemical separation, as the other radioactive members have half-lives that are either too long or too short.

Kulp *et al.* (1953) have determined the $^{206}Pb/^{210}Pb$ ages of ten pitchblende and uraninite samples, and have shown that they give very reliable ages over the range 60–1400 million years. However, Kohman and Saito (1954) have suggested that the method may not always give reliable ages for primary minerals, as a result of erratic radon loss, while secondary uranium minerals may not exhibit secular equilibrium between ^{238}U and ^{210}Pb. Ferrara, Ledent and Stauffer (1958) have used this method to date uranium-bearing sediments of the Western Alps. The uranium was fixed within the sediments during an interval of time between the upper Carboniferous and the middle Triassic. An average age of 151 million years suggests that the mineralization of the sediments occurred in early Permian time.

Lead minerals, cottunite ($PbCl_2$) and cannizzarite ($Pb + BiS$) from volcanic fumaroles, have been investigated by the ^{210}Pb method by Houtermans, Eberhardt and Ferrara (1963) and Eberhardt, Geiss, Houtermans, Buser and von Gunton (1955). In the cases of samples from Vesuvius, Italy, and the Valley of 10,000 Smokes, Alaska, all showed a high specific activity of ^{210}Pb, indicating a prior existence with either ^{226}Ra, ^{230}Th (Io) or ^{238}U. All the leads were anomalous, of the J Type (see p. 189) and gave ages of 360–450 million years. Similar anomalous leads from basic and ultrabasic rocks have been described by Tilton and Reed (1963), Tilton and Patterson (1956) and Marshall and Hess (1960). The highly anomalous ages are in accordance with the addition of radiogenic lead, not initially present in the basic magma, but possibly introduced by the assimilation or leaching from pre-existing rock.

The lead-210 method has found particular application to dating sediments, and for studies of mixing rates in oceans (Koczy, 1958) and the atmosphere (Maddock and Willis, 1961). In such studies, the precursor of lead-210, radon-222 provides the basis for its geochronological utility. Radon is continuously diffusing into the atmosphere, and its residence time in the atmosphere is governed by its half-life of $3 \cdot 8$ days. The total residence time is only a few weeks, and part is removed to the earth's surface by precipitation.

Lead-210 dating has been applied by Crozaz, Picciotto and de Breuck (1964) to studies of past precipitation rates in the Antarctic. Variation in the activity of ^{210}Pb in vertical firn samples are interpreted as showing that the initial ^{210}Pb concentration and the rate of water accumulation have remained constant for the past 100 years. From the decay curves, a rate of water accumulation of 6 ± 1 cm at the South Pole and 45 ± 3 cm at Base Roi Baudouin have been reported, and are in agreement with accumulation rates obtained by other methods.

Radiometric methods for investigating the Greenland Ice Cap have been described by Goldberg (1963).

6*

CHAPTER 8

RADIATION DAMAGE METHODS

I. Natural Fission Dating

Nuclear fission is a high-energy transmutation process that involves a re-arrangement of nucleons in the nucleus. While α-decay is predominant for elements above $Z=82$ (lead), at $Z=98$ (californium) the rate of decay by spontaneous fission is comparable to the α-decay rate. Spontaneous fission is a random process and for a particular nuclide can be assigned a half-life. When a fission event occurs, a range of fission products is produced with fission yields ranging from a small fraction to approximately 6%. In uranium minerals, an age can be determined by measuring the amount of an accumulated fission product and if the mineral retains the particular fission product, this will be proportional to its age, provided that the uranium content (parent) of the sample, the half-life for spontaneous fission and the fission yield (daughter) are known.

The use of spontaneous fission in uranium minerals as a means of dating was first used by Khlopin and Gerling (1947). The age of a sequence of uranium minerals was determined by means of the xenon/uranium ratio, assuming that the xenon is only produced through spontaneous nuclear fission. The rate of spontaneous fission in natural minerals is very slow, and it is difficult to measure the fission products with sufficient accuracy. Apart from xenon, krypton and strontium-90, other fission products have been measured in non-irradiated uranium (Kuroda, 1963). It has been known since 1947 (Khlopin and Gerling, 1947) that a proportion of the heavier isotopes of xenon in the earth's atmosphere have been produced by a nuclear fission process. Calculations based on the amount of xenon fission products from ^{238}U suggest an age for the earth that is too high compared with ages obtained by the K–Ar method. Contributions from the spontaneous fission of ^{235}U and ^{232}Th are negligible, but Kuroda (1961) has suggested that xenon may have been produced from ^{244}Pu, ^{247}Cm and ^{235}U neutron-induced fission.

II. Spontaneous Fission Track Dating

A relative newcomer to age studies, the method of spontaneous fission track dating described by Price and Walker (1962a), Fleischer and Price (1963a, 1964) and Maurette, Pellas and Walker (1964) holds much promise.

The passage of heavy particles through various media can be observed with the aid of a Wilson's cloud chamber or using photographic emulsion techniques. Noggle and Stiegler (1960) have observed that the passage of fission fragment particles through thin films of UO_2 gave rise to a population of tracks similar to those produced by the other techniques. Similarly spontaneous fission of uranium in natural silicates results in the formation of thin narrow tracks about 10μ in length (Silk and Barnes, 1959).

In natural mineral samples, uranium is the only element for which spontaneous fission is significant. The major contribution comes from the isotope ^{238}U, because of its high abundance relative to ^{235}U. Five other possible and generally insignificant sources of fission fragments are as follows.

(1) Spontaneous fission of ^{232}Th (Hyde, 1960) would only become important in thorium rich minerals; the concentration of thorium would have to exceed that of uranium by 10^5 for the track density to equal that for uranium fission. As the U/Th ratio for the common rock-forming minerals is generally low, fission track contribution from thorium would be negligible.

(2) Induced fission of heavy elements by terrestrial radiation could be induced by neutrons and γ-rays emitted in naturally occurring decay processes, e.g. neutrons produced by $(\alpha n),(\gamma n)$ reaction with beryllium.

(3) Cosmic ray fission induced by nucleons or mesons.

(4) Spallation recoils induced by cosmic rays.

(5) Cosmic ray induced fission of heavy element impurities.

Fission tracks produced in minerals can be observed by electron microscopy, but the techniques required are not amenable to routine study. Price and Walker (1962b, 1963) have rendered the observation of tracks easier by etching the mineral surface with appropriate reagents so as to enlarge the track "cylinder". In materials with a crystal lattice, the rough sides of the track cylinders are preferentially dissolved relative to the track-free mineral areas, such that the tracks could be observed with a conventional optical microscope. Apart from crystalline materials, Fleischer and Price (1963b) have observed fission tracks in natural glasses and tektites.

In general terms, the basic daughter–parent relation used in the age equations are as follows.

1. Parent.—The production of spontaneous fission tracks can be regarded as a chance event that can be described in terms of radioactive decay, in which case uranium constitutes the parent.

2. Daughter.—The daughter product of spontaneous fission is represented by the fission track.

3. Decay constant.—Some disagreement exists as to the real value for the decay constant for spontaneous fission. Fleischer and Price (1963a) have adopted an average value of 6.85×10^{-17} year^{-1} obtained by producing tracks in a synthetic mica placed against a uranium source for a known length of time, and by measuring track populations in minerals of known geological age.

If a mineral is irradiated by thermal neutrons, thermal fission of ^{235}U produces a new population of tracks which are related to the uranium concentrations as follows.

$$P_1 = N_v{}^{235}cR\phi \; \sigma \cos^2\theta_c$$

(Price and Walker, 1963), where P_1 is the induced number of tracks, N_v is the total number of atoms per unit volume, ^{235}c is the atomic concentration of ^{235}U, R is the mean track length, ϕ is the thermal neutron flux, σ is the fission cross-section and $\cos^2\theta_c$ is the critical angle for track production, mainly of use in the dating of glasses.

The term ϕ is measured by the simultaneous irradiation of uranium foil and counting the number of fission events by measuring ^{140}Ba or ^{99}Mo fission product abundances. In practice, the natural tracks are counted, after which the sample is irradiated and the uranium concentration determined from the new population. The induced tracks give a very accurate picture of the distribution of uranium, apart from being capable of measuring extremely low concentrations of uranium. Figure 39 illustrates some typical natural and artificially induced tracks in geological material.

The approximate geological age A is given by the equation:

$$A = p_s/N_v{}^{238}c\lambda_f R \; \cos^2\theta_c$$

(Fleischer and Price, 1963a), where p_s is the number of etched tracks per unit area of surface, λ_f is the decay constant for spontaneous fission of ^{238}U, p_s is the number of etched tracks per cm^2, N_v is the total number of atoms per unit volume, ^{238}c is the atomic concentration of ^{238}U, R is the mean length over which a fission fragment track can be revealed by etching and $\cos^2\theta_c$ is the critical angle for track etching.

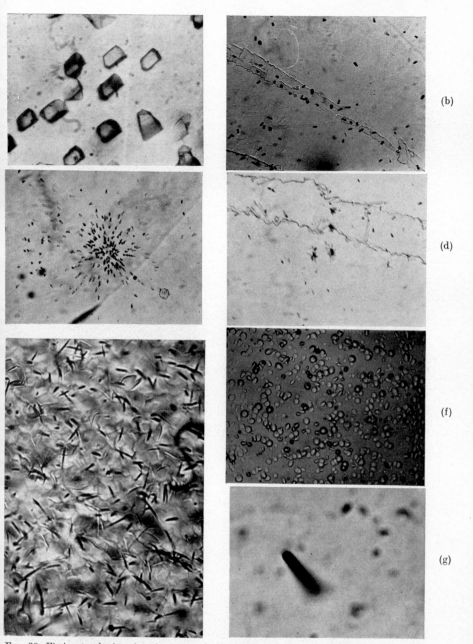

Fig. 39. Fission tracks in minerals. (a)–(f) are examples of *neutron induced fission* and (g) is an xample of natural fission. (a) Muscovite (the contoured margins of the pit indicate the angle of e track); (b) muscovite (higher concentration of uranium associated with a thin layer of biotite along 1 plane of muscovite); (c) lepidolite (local concentrations of uranium); (d) lepidolite (local point urces of uranium); (e) apatite (homogeneous distribution of uranium); (f) glass (the round character ' the tracks reflects the absence of crystal structure); (g) muscovite. The tracks in (a) are $\sim 2\mu$ in ameter, those in the other photographs are $\sim 10\mu$ in length. (Photographs, E. I. Hamilton.)

Depending upon the chemical composition of the sample, a suitable reagent is selected. Conditions for etching various minerals are given in Table 37.

TABLE 37. Conditions for etching minerals for fission track dating

Mineral	Etching media	Time	$T(°C)$
Quartz	KOH (aq. soln)	3 h	150
	48/HF	24 h	23
Orthoclase	48/HF	10 sec	23
Muscovite	48/HF	10–40 min	23
Biotite	48/HF	3–20 sec	23
Hornblende	48/HF	5–60 sec	23–60
Pyroxene	KOH (aq. soln)	1 min	220
Olivine	KOH (aq. soln)	8 min	220
	48/HF	5 sec	23
Apatite	HNO_2	10–30 sec	23
Calcite	10/HCl	50 sec	23
Garnet (pyrope)	KOH (aq. soln)	2 h	150
Zircon	KOH : NaOH(1 : 1)	10 sec	450

Table from Fleischer and Price (1964).

The optimum time for etching will result in a clean well defined track; if the etching is prolonged, the upper part of the track is dissolved, giving rise to a broad depression. Longer etching times are sometimes useful, particularly if the mineral section contains opaque inclusions or etch pits which might be confused for fission tracks. After etching, tracks are recognized as narrow black straight cylinders, randomly orientated, and are between 5μ and 20μ in length. In most minerals, the more concentrated track population after irradiation shows a general homogeneous distribution superimposed upon which are local point sources of tracks, high concentrations around radio-active minerals, or in particular areas such as mineral boundaries.

The age measurements given in Table 38 indicate that reliable ages can be obtained for a wide range of materials. The method is attractive in its simplicity as it does not require elaborate and expensive equipment, apart from the use of a nuclear reactor, which presents no problems. As the original tracks are very thin, they are rapidly destroyed by any subsequent heating. The relative rate of track fading at elevated temperatures varies for different minerals. The spontaneous fission age dates the last thermal event, and is probably more sensitive to this than K–Ar dating. When disagreement is observed between fission ages and those obtained by K–Ar,

Rb–Sr or Pb methods, the former are often lower in keeping with a thermal event that is not registered by the latter.

TABLE 38. Comparison of fission ages with those by the K–Ar, Rb–Sr methods

| Location | Fission age (million years) | Other methods | | U(ppm) by fission method |
		K–Ar ages	U(ppm)	
Tektites (Fleischer and Price, 1963a,b)				
Philippines	0·78	0·60–0·69	1·7–1·8	1·3
Indochina	0·41	0·56–0·66	1·6–1·7	1·3
Moldavia	12·6	13·5	1·8	2·1
Texas	35·3	34–35	1·2–1·5	1·2

Location	Mineral	Fission age (million years)	K–Ar, Rb–Sr, Pb, age (million years)
Minerals (Fleischer, Price, Symes and Miller, 1963)			
Texas (Llano)	Hornblende	1060±60	1075
Oregon (Taubenbeck)	Muscovite	108±27	95–100
Arizona		33±7	32·7
Colorado (Brown Co.)	Lepidolite	1390±250	1390
S. Rhodesia (Pope's Claim)	Lepidolite	670±100	2600–2700
S. Rhodesia (Salisbury)	Lepidolite	500±100	2600–2700
Minerals (Maurette *et al.*, 1964)			
Canada (Renfrew)	Muscovite	336	1000
Madagascar			
(Ampandrandara)	Phlogophite	190	533
(Malakialina)	Muscovite	575	514
Ceylon (Mahannea)	Phlogophite	338	485–570

$\lambda_f = 6\cdot89 \times 10^{-17}$.

While the fission method is in its infancy, its potential value should not be underestimated. Many of the low ages coincide with known thermal events in adjacent areas. It is to be hoped that systematic studies will be made in areas where the chronology is well established.

Recently, Fleischer, Naeser, Price, Walker and Marvin (1965) have dated zircons less than 1 million years to greater than 100 million years and obtained results that showed good agreement with ages obtained by isotopic methods. By selecting zircons with convenient concentrations of uranium, crystals weighing only a few micrograms have been successfully used.

A further use for this method will be studies of the rates of diffusion of uranium in various minerals. Apart from lattice diffusion, it should be possible to study the migration of uranium through minerals and along grain boundaries. The general approach would be to determine the distribution of induced tracks in a radioactive mineral, followed by heating, then subject it to further neutron bombardment and observe any changes in the distribution pattern.

Other possibilities that may be rewarding items of research are: (a) the determination of the uranium content of the sample after neutron irradiation requires a calibration of the neutron flux, and the use of an accurately calibrated radiation counter. Errors are possible in this measurement, and it is necessary to determine the neutron flux for each irradiation. Variations in the neutron flux recorded by the flux monitor may not be identical to those of the samples. This can be overcome by preparing a synthetic uranium standard by fusing a known amount of uranium in glass or silica to form a rod. Random thin slices can be cut from the rod, and the uranium content checked by neutron activation analysis. The degree of homogeneity of uranium in the glass can be monitored by α-counting. To determine the relation between neutron flux and production of tracks, a section of the glass standard is placed over the sample and, from the number of induced tracks formed, the uranium content of the sample can be determined; (b) by selective etching, it should be possible to observe track distributions and to date minerals contained in a normal thin section. Preliminary experiments by the author have shown that this is feasible for granites. Initially, dilute nitric acid is used to examine apatites, followed by hydrofluoric acid for biotites, feldspars and hornblendes; (c) examination of ore specimens, particularly galenas, by reflected light or carbon replicate techniques, which may be of interest in anomalous lead studies; (d) in uranium distribution studies, auto-radiographic methods are limited by the fading of the latent image; by increasing irradiation times, uranium distribution patterns and concentrations should be possible at extremely low concentrations.

It is far too early to evaluate the value of fission track dating and, while it will undoubtedly have limitations, it would appear to show more promise than other radiation methods.

III. Pleochroic Haloes

In thin sections of rocks, coloured pleochroic haloes are commonly found surrounding radioactive inclusions in biotite, fluorite and amphibole. The most common inclusions are zircon, xenotime, apatite and monazite, although in many instances they are too small, making a microscopic identification impossible. The coloration of the halo is a function of the α-radiation dose on the lattice of the host mineral. In detail, haloes contain a number of well defined, dark, concentric rings, the radii of which correspond to the ranges of particular α-particles. The range of particular α-particles varies in different minerals, depending on the density and stopping power (see p. 162).

The degree of halo coloration increases with radiation dosage until a saturation level is reached when the colour remains constant. Further irradiation can lead to colour inversion resulting in bleaching.

The α-particle radioactivity of the inclusion can be measured either by direct counting or by the nuclear emulsion method, while the intensity of the halo can be measured by photodensitometry. The specific α-particle activity required to produce a certain depth of halo darkening may be determined by bombarding mineral samples for a known length of time with a known flux of α-particles. In theory, once the specific α-particle activity is known, the degree of darkening should be proportional to time. However, various parameters cannot be measured with sufficient accuracy or precision to warrant the use of this method for dating.

Discussions of results and difficulties inherent to the method have been presented in papers by Joly (1907), Joly and Rutherford (1913) and, more recently, by Deutsch, Hirschberg and Picciotto (1956) and by Picciotto and Deutsch (1960).

Apart from dating, pleochroic haloes can provide other useful information. During a period of heating, the haloes are rapidly destroyed and the relation between K–Ar ages and loss of haloes may provide information regarding the extent of a thermal event.

If a radioactive inclusion contains thorium, its presence can be identified by the characteristic ring formed by ThC, which has a radius greater than the other α-emitters. Detailed analyses of halo radii in biotite show the presence in some samples of rings with a radius of 8·6 μ which have not been formed by the decay of the uranium or thorium series. Although not identified, they may be formed by the decay of ^{156}Gd or ^{146}Sm.

IV. Radiation Damage Method

Holland and Kulp (1950) suggested that measurement of radiation damage in minerals could be used as a method of age determination. Holland and Kulp (1954) and Hurley and Fairbairn (1957) have directed attention towards zircons. Apart from their widespread occurrence, they retain within the crystal lattice radiation damage as a result of prolonged α-particle bombardment. In order to apply this method to measuring geological time it is necessary that (a) the relation between radiation dosage and radiation damage must be constant since the mineral was formed; (b) the α-particle flux must have remained constant in the past.

The α dose rate is measured by thick source counting, and the resulting radiation damage is found by measuring the change in unit cell dimensions. The degree of radiation damage in zircons has also been correlated with change in birefringence and density. The age of a zircon is obtained from the ratio, dosage : activity. Highly radioactive zircons often tend to give a low age because of saturation radiation damage. Hurley and Fairbairn (1957) have shown that reliable ages can be obtained for zircons from granite, provided that surface activity and isolated internal point sources of uranium and thorium are eliminated by acid treatment. When low ages are obtained from low activity zircons, the ratio may indicate a later period of metamorphism.

While the method has provided interesting results, it is but rarely used. Two major sources of error are the heterogeneous distribution of uranium and thorium, and the annealing of radiation damage by heating during subsequent geological events.

V. Lead-alpha Method

The lead-alpha method, sometimes called the Larsen Method, has been extensively applied to accessory minerals, such as zircon, xenotime, monazite and thorite. In a closed chemical system, it is necessary to know the concentrations of parent, (uranium or thorium as determined by their α-particle radioactivity), daughter (total lead, assumed to be completely radiogenic in origin) and the rate of decay (λ) to calculate the geological age. The age will be meaningful only if the radioactive minerals crystallized contemporaneously with the major minerals, the lead is radiogenic only and has accumulated in the minerals since they were formed, and since their formation they have neither gained nor lost parent or daughter products. Isotopic analysis of lead separated from zircon and monazite suggests that in many cases the lead is predominantly radiogenic.

Zircon has proved to be the most suitable mineral for dating by this method. The total lead content is variable, ranging from 10–200 ppm, while the radioactivity is a direct function of the rock type. The method has proved to be fairly reliable when several minerals of different radioactivity and lead content can be used. Approximately 100 mg of sample is required, and a very careful separation of zircon from other accessory minerals is essential. Apatite, pyrite and "radioactive paint" on mineral surfaces can be removed by acid leaching without removing uranium, thorium and lead within the zircon lattice. In the case of low lead contents, precautions must be taken to reduce contamination. The results given in Table 39 show that paint (probably lead chromate) used on a magnetic separator can contain appreciable amounts of lead.

TABLE 39. Contamination by extraneous lead

Magnetic separator	Pb(ppm)
Brass track	1000
Vibrator paint	500
Magnet paint	10
Laboratory dust	1000 ppm

The results given in Table 40 indicate levels of radioactivity present in dust collected during a period of two weeks in the Department of Geology, Oxford. Measurements were made by a nuclear emulsion method and show a high level of radioactivity associated with a laboratory used for rock crushing and preparation of thin sections.

TABLE 40. Contamination of radioactive materials in dust

Top floor	
Photographic dark room	0·10
Nuclear emulsion laboratory	0·10
Middle floor	0·11
Ground floor	
Main hall	0·11
Slide preparations room	0·38
Rock crushing, cutting room	4·01

Values are α/cm^2 per h.

The total α-particle radioactivity is measured by direct counting, using either an activated zinc sulphide screen coupled to a photomultiplier or a proportional counter. For the latter, care must be taken to prevent the build up of surface static charge over the partially conducting silicate powder; this is conveniently accomplished by adding a conducting material, such as carbon, to the powder or by placing a fine metal grid over the sample.

Silicate materials will absorb part of the energy expended when an α-particle is emitted and the mean length of the α-particle "energy" should be greater than the absorption coefficient of the sample. This can be obtained by thin source counting, but is subject to many practical difficulties. It is far easier to use a thick sample and then to correct for absorption within the source. If the chemical composition and density of a mineral are known, the range of an α-particle in a particular matrix can be calculated by means of the Bragg and Kleeman formula (Hurley and Fairbairn, 1953):

$$\frac{\text{Range in A} \times \text{density of A}}{\text{Range in B} \times \text{density of B}} = \frac{\text{Atomic weight of A}}{\text{Atomic weight of B}}$$

The normal procedure is to compare the range in a mineral sample (A) with the α-particle range in standard air (B). The absorption correction can also be determined experimentally, if the total uranium and thorium content of the sample is known, by comparing the calculated activity with that measured from a thick source. Agreement between the calculated and measured α-activity indicates an accuracy for the α-activity measurements of about 5% (Gottfried, Senftle and Waring, 1956; Gottfried, Jaffe and Senftle, 1959). Taking into account the relative rate of α-particle emission from uranium and thorium, the theoretical thin source α-activity (in units of α-particles per mg per h) is given by

$$0 \cdot 3664\, U + 0 \cdot 0869\, Th$$

The total amount of lead is commonly measured by optical spectroscopy (Waring and Worthing, 1953), and is suitable over a concentration range of 0·5–3000 ppm Pb; in addition, standard colorimetric methods using dithizone are also widely used. Improved spectrochemical methods of Stern and Rose (1961), using standards having very similar chemical compositions to the samples, increase the accuracy of this method.

The basic age equations used for calculating the age of a sample are similar to those described by Keevil (1939). In the age range 0–200 million years an age is obtained from the formula

$$t = c \times \text{Pb(ppm)}\ \text{mg}^{-1}\,\text{h}^{-1} \tag{20}$$

c is related to the Th/U ratio; if uranium only is present, c is 2600 and, for thorium, is 1990. A Th/U ratio of about 1 is normally found for accessory zircon in igneous rocks, while the greatest variation is found in zircon from pegmatites. A large concentration in the Th/U ratio produces only a small error in the age and, for most zircons, a value of 2450 is chosen. If the uranium content and α-activity are known, the thorium content can be calculated with an accuracy of about $\pm 20\%$ from the equation

$$Th = \frac{\alpha - 0 \cdot 3664}{0 \cdot 0869}\ U\ \text{ppm} \tag{21}$$

where α is the α-activity per milligram per hour.

The equivalent uranium content of a sample may be estimated from its total α-activity according to the equation:

$$\text{eU} (\%) = 2{\cdot}75 \times 10^{-4} I \tag{22}$$

when I is the number of α-particles per milligram per hour.

If the age of the sample is older than 200 million years, a correction factor is required for the parent decay of uranium and thorium according to the equation:

$$t_0 = t - \tfrac{1}{2} K t^2 \tag{23}$$

where t is the age given by equation (20), and K is a decay constant based upon the Th/U ratios and equal to $1{\cdot}90 \times 10^{-4}$ if uranium only is present, and $0{\cdot}49 \times 10^{-4}$ if thorium only is present. The equations derived by Keevil involve several approximations, but are such that errors tend to compensate; for a more detailed account, see Gottfried, Jaffe and Senftle (1959).

Apart from experimental errors, and others as a result of natural leaching or addition of U, Th and Pb, the lead-alpha method gives acceptable ages which are equivalent to [206]Pb–[238]U ages. In cases of concordant [207]Pb–[206]Pb ages, lead-alpha ages are in close agreement, although the lead isotope age need not be equivalent to the geological age. Generally, good agreement is found between the lead-alpha age and the age obtained by Rb–Sr or K–Ar dating; although these ages may not mark a primary magmatic episode, they may indicate a late metamorphic event. Maximum disagreement occurs in Pre-Cambrian rocks, where the least lead-alpha ages tend to be lower. Some examples of lead-alpha ages from simple intrusions, and their comparison with other dating methods are given in Table 41 (Larsen and Keevil, 1947; Larsen, Keevil and Harrison, 1952; Quinn, Jaffe, Smith and Waring, 1957; Gottfried, et al., 1959).

The relative simplicity of determining lead-alpha ages, compared to other methods of dating, makes the method most attractive. However, the assessment of ages is very dependent upon the geochemical conditions present at the time of crystallization of the minerals from a magma, and subsequent events that may disturb the original geochemical pattern in the zircon–rock system. The lead-alpha method has its greatest potential as a reconnaissance tool, in which its speed and simplicity are particularly advantageous.

VI. Helium Dating

Helium is generated in radioactive minerals by the emission of α-particles in the decay of [238]U, [235]U and [232]Th, while negligible quantities are also produced by decay of such long lived natural radionuclides as [209]Bi, [142]Ce, [144]Nd, [147]Sm and [190]Pt. The quantity of helium produced in time t is

$$N_{\text{He}} = K_i (N_i^0 - N_i)$$

TABLE 41. Lead-alpha ages

Rock types	Locality	Mineral	Radioactivity $\alpha \, mg^{-1} \, h^{-1}$	Pb(ppm)	Lead-alpha age (million years)	Probable age	Reference
Quartz latite	San Juan, Colorado	Zircon	188 232	0·75 1·05	10 11	Miocene	Larsen and Cross (1956)
Monzonite	N. Mexico	Zircon	770	10	32	Oligocene	Gottfried et al. (1959)
Picrite	San Jose, California	Zircon	42	1·9	112	Early late Cretaceous	Gottfried et al. (1959)
Granodiorite	Sierra Nevada, California	Thorite Zircon	4670 400	205 15	88 93	Late Jurassic	Gottfried et al. (1959)
Granite	Rhode Island	Monazite Zircon Zircon	5494 190 300	585 19 25·5	220 243 208	Late or post Pennsylvanian	Quinn et al. (1952)
Gneiss	New Hampshire	Monazite Xenotime Zircon	2922 4053 242	470 530 30	342 266 297	Late Huronian	Gottfried et al. (1959)
Pegmatite	Routt Co. Colorado	Monazite Xenotime	1749 2390	1260 1555	1430 1420	Pre-Cambrian	Gottfried et al. (1959)

where N_{He} is the number of helium atoms produced from time t to the present, N_i^0 is the number of initial parent atoms at time t (^{238}U, ^{235}U, ^{232}Th), N_i is the number of initial parent atoms at the present time (^{238}U, ^{235}U, ^{232}Th), and K_i is the number of helium atoms produced per lead atom in each decay series :

i.e.
$$^{238}U \rightarrow {}^{206}Pb + 8He$$
$$^{235}U \rightarrow {}^{207}Pb + 7He$$
$$^{232}Th \rightarrow {}^{208}Pb + 6He$$

An age can be obtained by determining the total amount of helium, uranium ($^{238}U + {}^{235}U$) and thorium, together with a knowledge of the respective decay constants. In the early work (1908–28), helium was determined by chemical volumetric measurements, which were capable of detecting 10^{-10} ml, and measuring 10^{-9} ml He (Paneth and Peters, 1928). Damon and Green (1963) have shown that the early measurements gave low results compared to recent measurements by stable isotope dilution.

Uranium can be determined indirectly by measurement of the amount of radium present, provided that the sample is in radioactive equilibrium. The production rate of helium from uranium ($^{238}U = 0.9614$, $^{235}U = 0.0386$) is different from that of thorium ; while this is not important for young minerals, the failure to know the U/Th ratio leads to an error of 3.5% for 1000 million years and 11% for 2700 million years old samples. A measure of the total uranium plus thorium content of a sample can be obtained from the total α-particle activity (expressed in α-particles per mg per h), while the ratio U/Th can often be estimated for various rocks or minerals with sufficient accuracy.

Strutt (1905, 1908, 1909, 1910) has carried out extensive experiments to determine the amount, distribution and behaviour of helium in rocks and minerals. These results have shown that helium had accumulated in minerals with time, although many samples had obviously lost much of their helium. For this reason, the helium method was regarded as giving, at the best, only minimum ages.

Helium loss as a result of metamictization (Pabst, 1952) has been discussed by Hurley, Larsen and Gottfried (1956), Damon and Kulp (1957) and Holland and Gottfried (1955). Non-metamict zircons can retain much of their helium, while some metamict samples may have a helium retentivity as low as 1%. Anomalous ages can often be obtained because of the presence of superficial radioactivity in mineral surfaces and in cracks. If this is removed by dilute acid leaching, a mineral like magnetite, which retains helium, can give reliable ages ; similarly, fine-grained rocks, particularly basalts and dolerites, may shown a high degree of helium retentivity.

Fanale and Kulp (1962) have measured the helium, uranium and thorium content of magnetite and pyrite from the Cornwall Pennsylvania magnetite deposits associated with the Palisade Sill. When superficial uranium and thorium, not present in the actual lattice, were removed, the ores gave ages of 194 ± 4 million years, in agreement with a K–Ar age of 195 ± 5 million years on associated muscovite and 197 ± 5 million years for biotite. This supports the view of Keevil (1946) that nearly "perfect" minerals should show a high retention for helium.

In helium dating, it is assumed that at the time of formation the sample did not contain any helium. In general, this is true, but for some minerals, such as beryl, cordierite and tourmaline, it has been shown (Damon and Kulp, 1958; Gerling, 1961) that they can incorporate significant quantities of the inert gases. These minerals have a six-membered silicon tetrahedron forming ring structures enclosing large channels which may contain the excess gas while smaller amounts are probably incorporated within holes or defects that are present in most minerals.

The present status of the helium method (Damon and Green, 1963) indicates that reliable ages can be obtained provided that the sample has retained all its helium, and that the relation between the sites of helium and uranium atoms is known.

CHAPTER 9

COMMON LEAD METHOD

I. Introduction

At first sight, the relation between lead ores and the measurement of geological time would appear to be obscure. In lead isotope studies, three types of lead are recognized.

1. Primeval lead

It is assumed that, at some time during the early stages of the earth's history, a homogeneous distribution of lead isotopes existed such that lead from any area had the same isotopic composition. This initial distribution of lead isotope abundances was probably disturbed by the establishment of local variations in the U/Pb and U/Th ratios, as a result of chemical fractionation during cooling. These variations would result in different amounts of radiogenic lead being produced in different areas with time.

2. Common lead

Common lead differs from primeval lead by the addition of radiogenic lead through the radioactive decay of uranium and thorium. Of the four lead isotopes ^{204}Pb, ^{206}Pb, ^{207}Pb, ^{208}Pb, only that at mass 204 has not been produced in any significant quantities by a radioactive process since the earth solidified. Apart from galena, lead is found in varying amounts in other sulphides and also in the common rock-forming minerals, particularly potassium-rich feldspars. Lead ores are not associated with significant amounts of uranium and thorium; in the case of feldspars, the lead would be expected to be fairly evenly distributed throughout the crystal lattice, unlike uranium, which is concentrated at discrete points.

3. Radiogenic lead

Initially this lead is formed *in situ* by the radioactive decay of uranium or thorium. If a rock contains uranium and thorium, then the amount of radiogenic lead will increase with the passage of time. Subsequent remelting in a closed system would result in the mixing of radiogenic and common lead to produce a new common lead that would now contain greater quantities of ^{206}Pb, ^{207}Pb and ^{208}Pb than were originally present. If a very pure uranium or thorium deposit, devoid of any common lead, is "remobilized", then the subsequent stages of crystallization differentiation could lead to a deposit highly enriched in radiogenic lead. However, Sorensen (1963) has concluded that remobilization of ore bodies is an uncommon event.

Early chemical determinations of the atomic weight of lead from galena of varying geological age gave a consistent value of 207·21. Holmes (1937) has argued that galena ores could not be derived from sialic rocks, and must have a sub-sialic origin. Rock lead would consist of ore Pb + UPb + ThPb, hence its atomic weight would vary with geological time. In 1938, Nier determined, by mass spectrometry, the isotopic composition of lead from twelve lead ores for which chemical atomic weight measurements had been made. These results, given in Table 42, show that, although variations in the atomic weight were small, less than 0·02%, the isotopic composition varied considerably. Nier suggested that the two oldest leads may represent primeval lead, while the others may represent primeval lead that had become contaminated with radiogenic lead in the source regions where the ores were evolved. Later work by Wahl (1940) showed that Nier's suggestion of a primeval, original lead was not supported by analyses of old Fenno-Scandian leads, the isotopic variations for which were even greater than those found by Nier. Wahl proposed, from available evidence, that every lead ore body should have its own individual isotopic composition, which would depend upon how much uranium or thorium lead happened, through natural processes in the evolution of the ore, to come in contact with lead containing the 204 isotope.

Finally, a discussion of the work of Nier by Lane (1941), which includes further lead isotope analyses, suggests that basic rocks such as dunites, containing very small amounts of uranium and thorium, may in fact contain Nier's " original lead ". Also, if galenas of the same age have the same isotopic composition, then they have been formed from a homogeneous source region.

The distribution of lead in rocks and minerals has been described by Wedepohl (1956). The average lead content of micas and potassium feldspar (25 ppm) is greater than that for the lithosphere (1 ppm). The total lead content of the earth's crust has been increased by approximately 20% during the last 3×10^9 years by the addition of radiogenic lead. Among the major rock-forming minerals, the maximum enrichment in common lead occurs in potash feldspars, particularly those found in pegmatites. Lead replaces potassium in feldspars and, during magmatic differentiation, the ratio K/Pb increases, except for some extreme differentiates.

TABLE 42. Nier's original isotope analyses for lead ores

Mineral	Locality	Geological age (million years)	Isotopic abundances 204	206	207	208	Atomic weight Physical	Chemical
Galena	Great Bear Lake, Canada	Pre-Cambrian 1300	1·000	15·93	15·30	35·3	207·218	207·206
Galena	New South Wales, Australia	Pre-Cambrian	1·000	16·07	15·40	35·5	107·217	—
Cerusite	Broken Hill, N.S.W., Australia	Pre-Cambrian 950	1·000	15·92	15·30	35·30	207·217	207·210
Galena	Yancey Co., N.C., U.S.A.	Late Pre-Cambrian 600	1·000	15·93	15·28	35·20	207·216	207·209
			1·000	18·43	15·61	38·20	207·204	
Galena	Nassau, Germany	Carboniferous 240	1·000	18·10	15·57	37·85	207·206	207·210
Cerrusite	Eifel, Germany	Carboniferous 240	1·000	18·20	15·46	37·70	207·203	207·200
Galena I	Joplin, Missouri, U.S.A.	Late Carboniferous 230	1·000	21·65	15·88	40·80	207·178	207·220
II	Joplin, Missouri, U.S.A.	Late Carboniferous 230	1·000	21·60	15·73	40·30	207·175	—
			1·000	21·65	15·75	40·45	207·175	—
Galena	Metalline Falls, U.S.A.	Late Cretaceous 80	1·000	19·30	15·73	39·50	207·203	207·210
Cerrusite	Wallace, Idaho	Late Cretaceous 80	1·000	15·98	15·08	35·07	207·214	207·210
			1·000	16·10	15·13	35·45	207·217	
Wulfenite	Tucson Mts. Arizona, U.S.A.	Miocene 25	1·000	18·40	15·53	38·10	207·204	207·220
Vanadinite			1·000	17·34	15·47	37·45	207·215	—
Galena	Saxony, Germany		1·000	17·38	15·44	37·30	207·217	—

The distribution of lead in metamorphic rocks is hardly known, although Wedepohl (1956) has observed that an increase in lead content can occur when passing through the metamorphic grades. Variations in the isotopic composition of lead as a result of progressive metamorphism warrant detailed study.

The lead content of various rocks and minerals, as determined by Wedepohl, are given in Table 43.

TABLE 43. The distribution of lead in rocks and minerals (average values taken from Wedepohl, 1956)

Mineral	Lead content (ppm)	Rock	Lead content (ppm)
Chromite	15	Ultrabasic	3
Titanomagnetite	20	Gabbros	5
Olivine	1	Basalts	5
Pyroxene	6	Diabases	6
Amphibole	15	Quartzdiorite	8
Biotite	25	Granodiorite	15
Muscovite	15	Granite	20
Plagioclase	10	Adamellite	25
Sanidine	21	Rhyolites	30
K-Feldspar	27	Syenite	12
K-Feldspar (pegmatite)	100	Bauxite (variable)	7–140
Adularia	62	Phosphates	100
Zircon (variable)	10–300	Arkoses	5–30

While galena has figured and still does figure prominently in lead isotope studies, the range of types of material currently studied has increased to include lead in other sulphide minerals, and in the common rock-forming silicates. The isotopic analyses of lead at trace levels has necessitated the refinement of existing methods and the development of new techniques. The final lead isotope assay can be made either by solid- or by gas-source mass spectrometry.

II. Separation of Lead from Rocks and Minerals

In wet chemical procedures, the sample is dissolved in an appropriate solvent (e.g. HNO_3, $HF/HClO_4$, KHF, Na_2O_2) and lead is precipitated, together with common contaminants such as iron and aluminium, by heating with ammonia. The precipitate is dissolved in dilute nitric acid and process (i) or (ii) hereunder, or a combination of both, is used to separate lead in a chemically pure form. If contaminants (inorganic or organic) are present in the final lead salt, poor ion emission is obtained in the subsequent mass spectrometric analysis.

(i) Ammonium citrate is added to complex interfering elements, and the solution is adjusted to pH 9. Lead is then extracted into dithizone dissolved in chloroform. The lead dithionate is back extracted into dilute nitric acid, cyanide ions are added, the pH is adjusted to 9 and the lead extracted into chloroform; this serves to separate lead from zinc, both of which were extracted in the first stage. Lead is removed from the chloroform with dilute nitric acid and excess dithizone removed by shaking with chloroform. The solution containing the lead is evaporated to dryness, and any organic material is destroyed by heating with perchloric and nitric acids.

(ii) Contaminating ions can be removed by passing the sample solution in 1 N HCl through an ion exchange resin such as Dowex No. 1. Most contaminating ions are removed by elution with 1 N HCl, and the lead can then be removed from the resin with 8 N HCl (Kraus and Nelson, 1956; Cantanzaro and Kulp, 1964). If the samples contain major quantities of iron, this can be removed by a preliminary extraction from a 6 N HCl solution using diethyl ether.

Often ion exchange columns can be dispensed with and the resin added directly to a 1 N HCl solution of the sample in a beaker, followed by stirring and centrifuging-off the resin. The absorbed lead is then removed by the addition of 8 N HCl. This technique is advantageous in pre-concentrating lead from a solution, particularly in the presence of free colloidal sulphur, which impairs the flow of elutant through the columns.

Care should be exercised to prevent contaminating the sample with extraneous lead during the chemical processing, especially if microgram quantities of lead are to be separated. In a clean laboratory, with purified reagents, blanks of less than 1 μg can be achieved, although for samples containing a very small mounts of lead, such as basalts and meteorites, special laboratories are desirable that contain highly purified air and no metal components.

The extraction of lead from geological material has been described by Patterson, Goldberg and Inghram (1953), Tilton *et al.* (1955) and Chow and McKinney (1958).

Apart from wet chemical procedures, lead can be volatilized from minerals and rocks by direct heating. Marshall and Hess (1960) have heated samples to about 1500°C in a stream of nitrogen. The sublimates were condensed onto a water-cooled quartz finger, the lead removed with dilute nitric acid and then purified by a dithizone extraction. The separation of lead can be improved by heating under vacuum, and collecting the lead on a cold finger inserted into the furnace which may be of the induction or conventional resistance type. In this procedure, the amount of sublimed extraneous material is very small. Granite, basalts and most minerals require a temperature of about 1200°C over a period of about 8 h to quantitatively remove lead. Volatilization techniques favoured by Russian workers hold much promise as lead can be removed from very large sample weights without using large volumes of specially purified chemicals. A possible source of contaminating lead from the quartz furnace tube can occur during heating, but after rigorous chemical cleaning, the amount of lead likely to diffuse from deeper layers will be negligible. It is essential that either the lead yields are 100% or are representative of the isotopic composition of lead in the sample. In many samples, lead is likely to be present at two distinct sites: common lead distributed throughout the rock or mineral and radiogenic lead present at the sites of uranium and thorium atoms. Without the complete removal of lead, there will possibly be the tendency for the radiogenic lead to volatilize first. The completeness of chemical yield can be checked by comparing with the isotopic composition of lead obtained by conventional wet chemical methods. Alternatively, radioactive lead-210 can be incorporated into a resistant synthetic mineral phase, a known amount of which is added to the sample and the chemical yield determined by counting the activity of the sublimed lead. A further modification to the method in cases where bulky sublimates are formed is to introduce a stream of iodine vapour in contact with the sublimate at between 300 and 400°C. The extracted lead iodide vapour can then be removed in a cold trap.

High-vacuum volatilization techniques (Marshall, 1957; Marshall and Hess, 1961) would appear to hold much promise for more comprehensive analysis. Distinct and reproducible faint sublimate rings are produced along different sections of the cold finger, and undoubtedly other elements are being separated. It is possible that osmium can be separated by such techniques, and this would provide a very valuable method for concentrating the very small quantities of this element; other obvious choices for investigation would be zinc, mercury thallium and bismuth.

After the lead has been separated by any of these methods, it has to be converted to a form suitable for mass spectrometric measurement.

In solid source mass spectrometry, lead sulphide is generally preferred. The precipitation of a solution containing a few micrograms of sulphide can often be difficult. Apart from the actual precipitation, the purity and size of the sulphide granules is most important. If the precipitate is fine and is evenly spread over the surface of the filament, there is often a tendency for the deposit to evaporate, and the number of lead ions produced is very small; a coarser precipitate is often more suitable. In attempts to overcome the poor degree of ionization, various matrix modifiers are used, such as borax, ammonium nitrate or zirconia–silica mixtures (Zykov and Stupnikova, 1957; Akishin, Nikitin and Panchenkov, 1957). These are apparently successful, but the mechanism by which they influence ion production is not really known. Various filament materials, such as tantalum, tungsten and rhenium, are used either as single or triple assemblies. Further improvements may be made by the use of small furnace assemblies, where the lead sulphide is deposited on the inside of a small cylindrical furnace provided with a slit, which is covered by a heated tantalum filament. In such cases, the evaporated lead molecules make direct contact with the hot filament. Ehrenberg, Geiss and Taubert (1955) have described some furnace assemblies.

Apart from solid source mass spectrometry, the measurement of lead as tetramethyl ($PbMe_4$) gas is preferred by some workers. The lead tetramethyl (Bate, Miller and Kulp, 1951; Richards, 1962) method is mainly restricted to major lead minerals such as galena, although it can be applied, over a limited range of lead concentrations, to other minerals, such as pyrite, that contain trace quantities of lead (Ulrych, 1964). Generally, >50 mg of lead is required for the synthesis of $PbMe_4$.

If the galena contains impurities, it is first crushed and the impurities removed by heavy liquid separations. The purified galena is then dissolved in hydrochloric acid, followed by the addition of ethanol to precipitate the lead chloride, leaving most of the iron, zinc, copper and bismuth in the supernatant. The lead chloride is then reacted with MeMgI, which converts the lead to lead iodide and then to lead tetramethyl. Richards (1962) has suggested an alternative technique involving the reaction of LiMe in the presence of MeI, according to the equation:

$$3LiMe + MeI + PbCl_2 \rightarrow Me_4Pb + 2LiCl + LiI$$

This method has the advantage that conversion yields are greater and the reaction can be carried out at room temperature. The lead tetramethyl is then purified by distillation or by gas chromatographic methods.

Recently, Ulrych and Russell (1964) have described a free radical method for the synthesis of tetramethyl lead. Lead is distilled from about 4 g of pyrite at 900°C in a hydrogen furnace to form a lead mirror. Free alkyl radicals produced by the thermal decomposition of di-t-butyl peroxide (Benson, 1960) at 600°C react with the lead mirror and the resultant tetramethyl lead is trapped in a liquid air trap. A sample containing 500 μg of lead will produce 0·3 μl of tetramethyl lead which is then purified by gas–liquid chromatography.

The mass spectrum of lead tetramethyl consists of positive ions of lead, lead hydride, lead monoethyl, dimethyl, trimethyl and tetramethyl. The trimethyl peaks are the most abundant and are found in the range mass 248–255. The principal peaks at masses 249, 251, 252, 253 correspond to the lead isotopes 204, 206, 207, 208 in association with three methyl groups. Other peaks result from the occurrence of carbon-13 in the ion and from loss of a single hydrogen atom from the lead trimethyl ion. A small peak at mass 255 is caused by the combination of two carbon-13 atoms with lead-208 in lead trimethyl. Peaks are also observed due to lead and lead hydride ions. In addition, mercury contamination from diffusion pumps can result in a background at the critical mass of 249.

The lead tetramethyl spectrum is clearly complicated, and it is often necessary to correct the observed lead spectrum for other superimposed peaks caused by the presence of hydrocarbon impurities. In a gas mass spectrometer, the source conditions for ionization are fairly constant and can be more or less reproduced for different samples, in contrast to the emission of ions from solids which cannot be reproduced. Gas analysis offers a distinct advantage in that samples and standards can be analysed alternately. Isotope fractionation occurs in lead tetramethyl analysis; as the gas enters the ionization chamber through a fine leak, the gas flow is viscous and fractionation of istopes will occur. While the precision of isotope measurement is greater than that obtained from solids, tetramethyl lead analysis is at present restricted to samples containing at least 100 ppm of lead, although it may be extended by extracting sufficient lead from samples containing smaller quantities of lead by volatilization techniques.

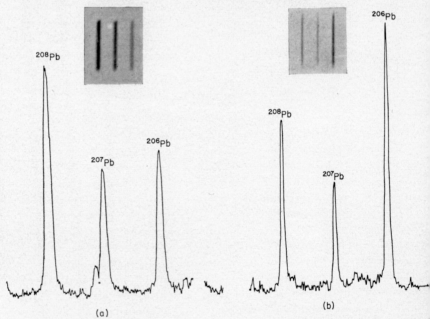

Fig. 40. Mass spectra for (a) common lead and (b) radiogenic ^{206}Pb samples obtained with an A.E.I. MS7 spark source mass spectrograph. (Photographs by permission of A.E.I.)

In lead isotope studies, it is the very small variations that become significant, and it is essential to be able to measure isotope ratios with both high precision and accuracy. Errors are frequently made when determining the least abundant lead isotope at mass 204.

Instrument errors can be assessed by using an artificially produced lead isotope standard. Accurately known weights of highly enriched ^{206}Pb and ^{208}Pb are mixed in known proportions and the calculated ratios are compared with those measured. The isotopic purity of the two standard isotopes would have to be determined by mass spectrometry, but any errors introduced by this measurement would be small. It is most important that the standard lead be chemically pure in terms of both cations and anions. In measuring lead isotope ratios from solids, a precision of a few tenths of a per cent is obtained by most laboratories, but to reduce this to 0·1% or less often necessitates modifications to conventional spectrometers.

Not very much value can be placed upon a single lead isotope analysis from one locality, and it is only when many samples are measured that lead isotope ratios can be properly interpreted.

The mass spectrograph can be used to determine lead isotope abundances, but with existing techniques it is only possible to determine the abundances of lead-206, -207 and -208; the emission of lead at mass 204 is too weak to measure with comparable accuracy. The spectra of samples of common and radiogenic lead are given in Fig. 40 together with line density plots. At present, the mass spectrograph would appear to be most useful in reconnaissance work, although it is to be hoped that improvements in recording techniques will enable ^{204}Pb to be measured with sufficient accuracy.

III. The Interpretation of Lead Isotope Abundances

A. MODELS

In all lead isotope abundance studies two assumptions are made. (1) At an early stage in the cooling of the earth all lead had the same isotopic composition; this lead has been called *primordial* or *primeval* lead. (2) Any variations that have subsequently occurred in lead isotope ratios are due solely to the addition of radiogenic lead that has been added to primeval lead.

Since the time that the earth was formed, 50% of ^{238}U has decayed to ^{206}Pb, and 90% of ^{235}U has decayed to ^{207}Pb. After an early rapid growth stage, the growth curve for ^{207}Pb is now asymptotically approaching a limiting value, while the former growth curve is more linear with time.

Consider a unit area of crust or mantle containing primordial lead together with uranium and thorium. Very early in the earth's cooling history, that may be identified as the age of the earth, no radiogenic lead would have been formed from the uranium and thorium. With the passage of time, radiogenic ^{206}Pb, ^{207}Pb and ^{208}Pb would grow at the sites of uranium and thorium

7

atoms. The growth of radiogenic lead from a constant U/Pb, Th/Pb environment would be expected to follow a simple growth curve, illustrating the
increase of daughter isotopes with time. This is analogous to a decay curve,
which would show the decrease in parent isotopes with time. After sufficient
time has passed, it is assumed that geological processes have occurred resulting in the formation of an ore solution. If all the lead in the unit area were
extracted to form the potential ore deposit, the common lead so formed
would consist of primordial lead plus radiogenic lead, which would have
been isolated from uranium and thorium as a result of differences in geochemical behaviour. The process of ore formation has therefore "sampled"
the isotopic composition of lead that has evolved to some point along the
lead growth curve. When the ore solution is emplaced as an ore deposit, the
lead isotope abundances developed in the source area at the time of mineralization are preserved, as a lead mineral like galena would not contain any
significant quantities of either uranium or thorium. Common lead formed by

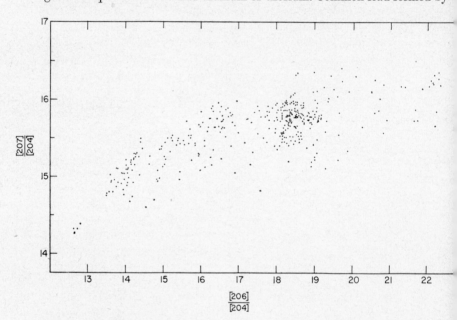

Fig. 41. ^{207}Pb/^{204}Pb against ^{206}Pb/^{204}Pb diagram of common leads of different origin. (After
Houtermans and Eberhardt 1960).

the removal of lead at a particular time from a constant U/Pb, Th/Pb environment is referred to as single stage lead, and can be described in simple
mathematical terms. On the other hand, if a potential single stage lead passes
through crustal rocks and is contaminated by the addition of radiogenic
lead that has been produced in a different U/Pb, Th/Pb system, the resulting

lead is termed anomalous. If the age of the lead mineral is calculated according to a single stage growth curve, then the excess radiogenic lead will result in an age younger than the true age, i.e. the amount of radiogenic lead present will be greater than could have been formed since the original single stage lead was formed, prior to it becoming contaminated by the addition of excess crustal lead.

In Fig. 41 a number of common leads of different origin are illustrated by plotting the ratio $^{207}Pb/^{204}Pb$ against $^{206}Pb/^{204}Pb$. It is apparent that the various points are distributed about a curve for which old lead minerals correspond to low and younger minerals to high lead isotope ratios. Similarly, the distribution of ordinary leads given in Russell and Farquhar (1960a), and excluding samples that are accepted as being anomalous, is shown in Fig. 42. The points conform within narrow limits to a single stage growth curve, and the small scatter of points may be caused by (a) differences in the U/Pb ratio for the source area in which the leads spent the first part of their history; (b) some of the points may in fact represent anomalous leads that have not been recognized; (c) errors in the measurement of lead isotope ratios, in particular, the abundance of ^{204}Pb.

The growth of radiogenic lead with time can be expressed as follows.

$$\text{Number of U, Th atoms at } t_0 = \begin{cases} ^{238}U\ e^{\lambda t_0} \\ ^{235}U\ e^{\lambda' t_0} \\ ^{232}Th\ e^{\lambda'' t} \end{cases}$$

$$\text{Number of U, Th atoms at } t_m = \begin{cases} ^{238}U\ e^{\lambda t_m} \\ ^{235}U\ e^{\lambda' t_m} \\ ^{232}Th\ e^{\lambda'' t_m} \end{cases}$$

where t_0 is the age of earth and t_m is the time of mineralization. λ is the decay constant for ^{238}U, λ' for ^{235}U and λ'' for ^{232}Th.

Equations (24)–(26) depict the number of parent uranium atoms that have decayed to daughter lead atoms during the period t_0–t_m:

$$^{238}U\ e^{\lambda t_0} - {}^{238}U\ e^{\lambda t_m} \tag{24}$$

$$^{235}U\ e^{\lambda' t_0} - {}^{235}U\ e^{\lambda' t_m} \tag{25}$$

$$^{232}Th\ e^{\lambda'' t_0} - {}^{232}Th\ e^{\lambda'' t_m} \tag{26}$$

Therefore the number of daughter lead atoms at time t_m, illustrated by the decay $^{238}U \rightarrow {}^{204}Pb$, will be:

$$(^{206}Pb)_{t_m} = (^{206}Pb)_{t_0} + (^{238}U)_{t_0} - (^{238}U)_{t_m} \tag{27}$$

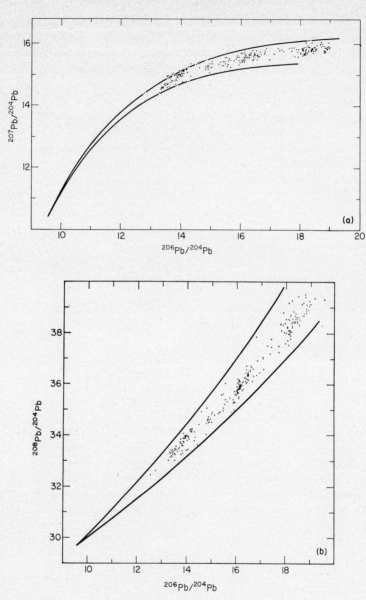

FIG. 42. Plots of (a) $^{207}Pb/^{204}Pb$ against $^{206}Pb/^{204}Pb$ and (b) $^{208}Pb/^{204}Pb$ against $^{206}Pb/^{204}Pb$ for ordinary leads.   (Data taken from Russell and Farquhar, 1960a.)

where ^{238}U is the present day uranium content. Dividing (27) through by the non-radiogenic ^{204}Pb, we have

$$\left(\frac{^{206}Pb}{^{204}Pb}\right)_{t_m} = \left(\frac{^{206}Pb}{^{204}Pb}\right)_{t_0} + \left(\frac{^{238}U}{^{204}Pb}\right)(e^{\lambda t_0} - e^{\lambda t})^m \tag{28}$$

Similar equations, which are one form of the growth curve, may be written for ^{207}Pb and ^{208}Pb. By combining the equations for ^{206}Pb and ^{207}Pb so as to eliminate the present day $^{235}U/^{204}Pb$ ratio, we obtain (29), the equation of a straight line having a slope S given by

$$S = \frac{\left(\dfrac{^{207}Pb}{^{204}Pb}\right)_{t_m} - \left(\dfrac{^{207}Pb}{^{204}Pb}\right)_{t_0}}{\left(\dfrac{^{206}Pb}{^{204}Pb}\right)_{t_m} - \left(\dfrac{^{206}Pb}{^{204}Pb}\right)_{t_0}} = \frac{e^{\lambda' t_0} - e^{\lambda' t_m}}{137 \cdot 8(e^{\lambda t_0} - e^{\lambda t_m})} \tag{29}$$

$^{238}U/^{204}Pb$ today $= 137 \cdot 8$.

Equation (29) is known as the Houtermans "isochron equation" and, if the measured $^{207}Pb/^{204}Pb$ ratios are plotted against the $^{206}Pb/^{204}Pb$ ratio, all leads of the same age will lie along a straight line that passes through the primeval abundances for lead $^{206}Pb/^{204}Pb$ and $^{207}Pb/^{204}Pb$. This method of interpreting lead isotope abundances was first described by Gerling (1942), Holmes (1946) and Houtermans (1946) and is known as the Holmes-Houtermans (H.H.) model. The graphical form of the H.H. model for common leads is given in Fig. 43, from which model ages may be determined from the intersection of the two measured uranium/lead ratios. An alternative method is to use tables prepared from calculations carried out by a computer as given in Table 44. The curved lead development lines are described by equation (28) while the fan of isochrons for different model ages along which leads of the same age will lie is given by equation (29). The isochrons cut the growth curves, and therefore samples of the same age are independent of the geochemical values U/Pb, Th/U. The equations for calculating these ratios in the source of the ores from the measured abundances are as follows.

$$\frac{^{238}U}{^{204}Pb} = \frac{\left(\dfrac{^{206}Pb}{^{204}Pb}\right)_{t_m} - \left(\dfrac{^{206}Pb}{^{204}Pb}\right)_{t_0}}{e^{\lambda t_0} - e^{\lambda t_m}} \tag{30}$$

$$\frac{^{232}Th}{^{238}U} = \frac{\left(\dfrac{^{208}Pb}{^{204}Pb}\right)_{t_m} - \left(\dfrac{^{208}Pb}{^{204}Pb}\right)_{t_0}}{\left(\dfrac{^{206}Pb}{^{204}Pb}\right)_{t_m} - \left(\dfrac{^{206}Pb}{^{204}Pb}\right)_{t_0}} \frac{e^{\lambda t_0} - e^{\lambda t_m}}{e^{\lambda'' t_0} - e^{\lambda'' t_m}} \tag{31}$$

where λ'' is the decay constant for thorium-232. The calculated ratios show a very small spread and suggest that the U/Pb and Th/U for the source regions are fairly constant. In the H.H. model, it is assumed that small initial differences of a regional nature exist in the geochemical ratios as a result of early chemical fractionation; however, since t_0, these ratios have not changed except by the radioactive decay of uranium and thorium to lead.

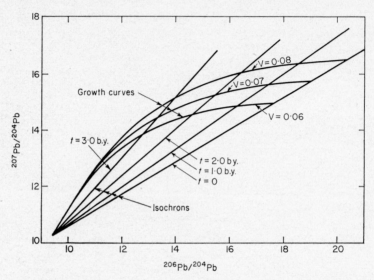

Fig. 43. Holmes-Houtermans model illustrating the relation between isochrons and "lead development" lines ($V = {}^{238}U{}^{204}Pb$). (After Russell and Farquhar, 1960.)

Another approach to lead isotope studies, developed by Allan, Farquhar and Russell (1953), Collins, Russell and Farquhar (1953), Collins, Farquhar, and Russell (1954), Russell, Farquhar, Cumming and Wilson (1954), Russell and Farquhar (1960a), Russell (1963) and Ostic (1963), is referred to as the Russell, Farquhar, Cumming (R.F.C.) model.

The measurement of a large number of lead isotope ratios clearly shows that all single stage common leads are closely distributed about a single growth curve. When ratios subject to errors in ${}^{204}Pb$ measurement and leads that have been passed through more than one stage are rejected, the fit to a single curve is remarkably good. The R.F.C. model is based upon Alpher and Herman's (1951) interpretation of Nier's original analyses of twelve galenas. Essentially, they state that in the early history of the earth there was a complete mixing of uranium, thorium and lead isotopes. The most obvious source region would be provided by mantle material, although a well mixed crust would also meet the requirements. Since the initial mixing stage, all leads

TABLE 44. Table for dating ordinary leads using the Holmes-Houtermans equation

ϕ	t	ϕ	t	ϕ	t	ϕ	t
0·5946	0	0·6508	600	0·7229	1200	0·8171	1800
0·5962	20	0·6529	620	0·7257	1220	0·8207	1820
0·5979	40	0·6550	640	0·7285	1240	0·8243	1840
0·5996	60	0·6572	660	0·7313	1260	0·8280	1860
0·6013	80	0·6594	680	0·7341	1280	0·8317	1880
0·6030	100	0·6616	700	0·7370	1300	0·8354	1900
0·6047	120	0·6638	720	0·7399	1320	0·8392	1920
0·6065	140	0·6660	740	0·7428	1340	0·8430	1940
0·6083	160	0·6683	760	0·7457	1360	0·8468	1960
0·6101	180	0·6706	780	0·7486	1380	0·8507	1980
0·6119	200	0·6729	800	0·7516	1400	0·8546	2000
0·6137	220	0·6752	820	0·7546	1420	0·8586	2020
0·6155	240	0·6775	840	0·7576	1440	0·8626	2040
0·6173	260	0·6798	860	0·7606	1460	0·8666	2060
0·6191	280	0·6822	880	0·7637	1480	0·8707	2080
0·6210	300	0·6846	900	0·7668	1500	0·8748	2100
0·6229	320	0·6870	920	0·7699	1520	0·8789	2120
0·6248	340	0·6894	940	0·7731	1540	0·8831	2140
0·6267	360	0·6918	960	0·7763	1560	0·8873	2160
0·6286	380	0·6943	980	0·7795	1580	0·8916	2180
0·6305	400	0·6968	1000	0·7828	1600	0·8959	2200
0·6324	420	0·6993	1020	0·7861	1620	0·9003	2220
0·6344	440	0·7018	1040	0·7894	1640	0·9047	2240
0·6364	460	0·7044	1060	0·7927	1660	0·9092	2260
0·6384	480	0·7070	1080	0·7961	1680	0·9137	2280
0·6404	500	0·7096	1100	0·7995	1700	0·9183	2300
0·6424	520	0·7122	1120	0·8029	1720	0·9229	2320
0·6445	540	0·7148	1140	0·8064	1740	0·9276	2340
0·6466	560	0·7175	1160	0·8099	1760	0·9323	2360
0·6487	580	0·7202	1180	0·8135	1780	0·9370	2380
0·9417	2400	0·9921	2600	1·0485	2800	1·1100	3000
0·9465	2420	0·9975	2620	1·0543	2820	1·1166	3020
0·9513	2440	1·0030	2640	1·0602	2840	1·1232	3040
0·9562	2460	1·0085	2660	1·0661	2860	1·1299	3060
0·9611	2480	1·0141	2680	1·0721	2880	1·1366	3080
0·9661	2500	1·0197	2700	1·0782	2900	1·1434	3100
0·9712	2520	1·0254	2720	1·0844	2920	1·1503	3120
0·9764	2540	1·0311	2740	1·0907	2940	1·1573	3140
0·9816	2560	1·0369	2760	1·0971	2960	1·1644	3160
0·9868	2580	1·0427	2780	1·1035	2980	1·1716	3180
1·1788	3200	1·2357	3350	1·2969	3500		
1·1971	3250	1·2556	3400	1·3182	3550		
1·2159	3300	1·2760	3450	1·3409	3600		

$$* \phi = \frac{\left(\dfrac{^{207}\text{Pb}}{^{204}\text{Pb}}\right)_{t_m} - \left(\dfrac{^{207}\text{Pb}}{^{204}\text{Pb}}\right)_{t_0}}{\left(\dfrac{^{206}\text{Pb}}{^{204}\text{Pb}}\right)_{t_m} - \left(\dfrac{^{206}\text{Pb}}{^{204}\text{Pb}}\right)_{t_0}} = \frac{(e^{\lambda' t_0} - e^{\lambda' t_m})}{137 \cdot 8(e^{\lambda' t_0} - e^{\lambda'' t_m})}$$

where $t_0 = 4560$ million years. (Reproduced with kind permission of E. R. Kanasewich.)

have developed in a constant U/Pb, Th/U environment along a single growth curve. However, the geochemical ratios and the growth curve represent average values, and hence both parameters would be expected to show a limited range of values.

Because of this, ages obtained by the R.F.C. model have rather large errors (Russell *et al.*, 1954), a point often ignored when comparing with lead ages obtained from other models. The R.F.C. method will yield two independent model ages for a single sample, namely from the $^{206}Pb/^{204}Pb$ and $^{208}Pb/^{204}Pb$ ratios (equation(32)).

Use of the $^{208}Pb/^{204}Pb$ ratio was first investigated by Collins *et al.* (1953), but often the model ages do not agree with those obtained from uranium leads. The half-life of ^{232}Th is about $1\cdot4 \times 10^9$ years; while the primeval abundance makes up three-quarters of the total ^{208}Pb content. For this reason, very accurate measurements are needed to determine the small addition of radiogenic thorium lead. In many lead deposits, the $^{232}Th/^{204}PB$ ratios show greater variations than those of $^{238}U/^{204}Pb$. This is undoubtedly due to differences in chemical behaviour of uranium and thorium relative to lead.

The assumption of a homogeneous source for leads results directly to the equations:

$$\left(\frac{^{206}Pb}{^{204}Pb}\right)_{t_m} = \left(\frac{^{206}Pb}{^{204}Pb}\right)_{t_0} - 137\cdot8\,\frac{^{238}U}{^{204}Pb}\,(e^{\lambda t_0} - e^{\lambda t_m})$$

$$\left(\frac{^{207}Pb}{^{204}Pb}\right)_{t_m} = \left(\frac{^{207}Pb}{^{204}Pb}\right)_{t_0} - \frac{^{235}U}{^{204}Pb}\,(e^{\lambda' t_0} - e^{\lambda' t_m}) \qquad (32)$$

$$\left(\frac{^{208}Pb}{^{204}Pb}\right)_{t_m} = \left(\frac{^{208}Pb}{^{204}Pb}\right)_{t_0} - \frac{^{232}Th}{^{204}Pb}\,(e^{\lambda'' t_0} - e^{\lambda'' t_m})$$

Once the primeval abundances and geochemical ratios have been determined, these equations are sufficient to define uniquely the isotope ratios in the homogeneous source at any time t. While the assumptions of the H.H. and R.F.C. models are quite different, the mathematical theory is identical. Both models will give almost identical ages, provided that the same constants (age of earth, primeval abundance, etc.) are used; the R.F.C. growth curve model can give meaningful results only if the isotope ratios conform closely to a single growth curve.

Lead isotope studies are being directed towards a means of distinguishing between ordinary and anomalous leads. The lead isotope ratios of the former would appear to fit simple mathematical models, while the latter do not, and contain excess amounts of radiogenic lead.

Stanton (1955a,b) has suggested criteria for recognizing ordinary leads, the method for which has been described by Stanton and Russell (1959), and is known as the Russell, Stanton, Farquhar (R.S.F.) method, and by Russell and Farquhar (1960a). The R.S.F. method is a special case of the R.F.C. model applied to conformable ore deposits. Such deposits, first recognized by King and Thompson (1953), are generally associated with cataclastic volcanic rocks, have been deposited in near-shore sediments, and show a very variable relation to structure. Many conformable deposits have been interpreted as hydrothermal replacement deposits.

Stanton has proposed that conformable Pb–Zn–Cu deposits are formed from lead, together with other ore-forming elements, emitted from volcanoes and later trapped in near-shore sediments. Subsequent action by sulphur bacteria in a reducing environment gives rise to lenses and pockets consisting predominantly of iron sulphide. Later compaction of the volcanic ash beds removes water-soluble lead minerals, such as cotunnite, which pass upwards and replace iron in the sulphide lenses to give rise to secondary conformable lead deposits. Oceanic volcanoes emitting basalts would be expected to represent average mantle material, although specific rock types may be produced by crystal fractionation and chemical differentiation. The primary magmas do not have to pass through any crustal rocks, and consequently would not be contaminated by the addition of radiogenic leads. Furthermore, the source region may be expected to have a uniform distribution of Pb, U and Th. A conformable lead will therefore be, to a first approximation, a lead representative of the upper mantle, or perhaps a deeper zone.

If the mantle is well mixed, all conformable leads should have $^{206}Pb/^{204}Pb$, $^{207}Pb/^{204}Pb$ ratios that lie along a single growth curve, with meteorite lead at one end and modern lead at the other. The distribution of points about a primary growth curve when plotting $^{207}Pb/^{204}Pb$ against $^{206}Pb/^{204}Pb$ for many lead ores, given in Fig. 42, show a general scatter of points, but, as shown in Fig. 44 (Stanton and Russell, 1959), if conformable lead deposits are chosen, all points lie on a single growth line.

Stanton's theory implies that, if a potential lead ore passes through, or is intruded into, pre-existing rocks, the assimilation or extraction of pre-existing lead will give rise to an anomalous lead ore. The degree of anomaly in an epigenetic ore deposit will depend upon many factors such as (a) the thickness of rock traversed by the initially conformable type lead; (b) the time taken in transit before the ores crystallized and exchange between U, Th, Pb was terminated and (c) the relative amount of primary lead to crustal lead; the U/Pb ratio and pre-mineralization age of the rocks traversed by primary lead ores.

In its simple form, the Stanton theory is not compatible with geological evidence as lead ores are not associated with primary basalt terrains, but in its wider regional application it would appear to be valid. Ostic (1963) has shown that the geological criteria used by Stanton can be used to identify

7*

conformable lead deposits when the isotopic composition is very uniform throughout the deposit. However, in some instances, conformable leads can significantly anomalous. If a conformable deposit has originated from mantle material, it would be reasonable to expect primary mantle leads, but it is

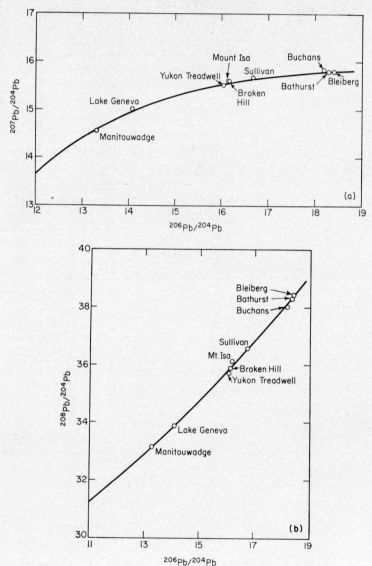

FIG. 44. The relationship between $^{207}Pb/^{204}Pb$ against $^{206}Pb/^{204}Pb$ and $^{208}Pb/^{204}Pb$ against $^{206}Pb/^{204}Pb$ for *ordinary* common leads as defined by the parameters of Russell, Stanton and Farquhar for *conformable* leads. (After Russell and Farquhar, 1960a.)

equally possible for a conformable deposit to be formed by crustal processes. It is also possible for apparently deep-seated lead to be contaminated in passing through pre-existing rocks containing radiogenic lead. Houtermans, Eberhardt and Ferrara (1963) have criticized Stanton's hypothesis on the grounds that the isotopic composition of lead from basic, ultrabasic (Tilton and Patterson, 1956; Marshall and Hess, 1960; Tilton and Reed, 1963; Patterson and Duffield, 1963) and volcanic fumaroles (Houtermans, 1960; Houtermans, Ferrara and Eberhardt, 1963) are often anomalous and contain excess radiogenic lead. Many of these samples intrude crustal rocks and contamination by radiogenic lead leached from country rocks, possibly by heating, would be expected. In the case of oceanic basalts, such as Hawaii, such an explanation would not be plausible. In some of the alkali differentiates of Hawaii about 2 ppm of uranium (Heier, McDougall and Adams, 1964; Hamilton, 1964) and about 7 ppm of thorium (Heier et al., 1964) have been recorded. While these rocks are less than 10 million years old, they illustrate that oceanic basalt magma, when differentiated, is capable of producing rocks containing uranium and thorium in amounts comparable to those found in continental rocks. The possibility exists that U- and Th-rich rocks formed by chemical differentiation of basic magma may exist at other deeper horizons. While such rocks would be only small in volume, their lateral dimensions may be considerable. Having crystallized at lower temperatures than the more abundant basic types, if reheated during later volcanic episodes, they would be the first to be remobilized and would be expected to be enriched in uranium, thorium and radiogenic lead.

The observed uranium content of oceanic rocks may not represent the uranium content of the corresponding magma. If lead is capable of being emitted from active volcanic sites in sufficient quantity to form conformable lead deposits, uranium must surely be emitted in comparable or perhaps greater amounts. While lead can be precipitated with relative ease, uranium as U^{6+} is likely to be extremely mobile and would not now be found co-existing with lead.

Presuming that both uranium and thorium have been concentrated by magmatic differentiation, it is possible that at depth a combination of "mantle outgassing" and chemical differentiation has lead to local zones that contain abnormally high concentrations of these elements. During a volcanic extrusive episode, uranium at least would tend to migrate towards the active extrusive site and, in so doing, the " released" radiogenic lead would follow. More information is needed before Stanton's hypothesis can be dismissed, in particular the amount of lead, uranium and thorium emitted during volcanic eruptions.

Recently, Patterson (1964) has described a mechanism for lead isotope evolution on a continental scale in the earth that involves the continuous transportation of uranium from the interior of the earth to an outer proto-continental layer throughout geologic time.

B. Constants Used in the Models

Different symbols used by the Holmes-Houtermans, and Russell schools of lead isotope studies are given in Table 45.

Table 45. Symbols commonly used in lead isotope abundance studies

Isotope ratio	Symbols used by Russell, Farquhar and others			Symbols used by Houtermans, Holmes and others		
	Present $(t=0)$	Time t	Primeval $(t=t_0)$	Present $(p=0)$	Time p	Primeval $(p=W)$
$^{206}Pb/^{204}Pb$	a	x	a_0	α		αw
$^{207}Pb/^{204}Pb$	b	y	b_0	β		βw
$^{208}Pb/^{204}Pb$	c	z	c_0	γ		γw
$^{238}U/^{204}Pb$	αV	$\alpha Ve^{\lambda t}$	$\alpha Ve^{\lambda t_0}$	μ	$\mu e^{\lambda p}$	$\mu e^{\lambda w}$
$^{235}U/^{204}Pb$	V	$Ve^{\lambda' t}$	$Ve^{\lambda' t_0}$	$\epsilon\mu$	$\epsilon\mu e^{\lambda' p}$	$\epsilon\mu e^{\lambda' w}$
$^{232}Th/^{204}Pb$	W	$We^{\lambda'' t}$	$We^{\lambda'' t_0}$	$\kappa\mu$	$\kappa\mu e^{\lambda'' p}$	$\kappa\mu e^{\lambda'' w}$

Physical constants : $1/\epsilon = \alpha = 137\cdot8$

$$\lambda = 0\cdot1537 \times 10^{-9} \text{ year}^{-1}$$
$$\lambda' = 0\cdot9722 \times 10^{-9} \text{ year}^{-1}$$
$$\lambda'' = 0\cdot0499 \times 10^{-9} \text{ year}^{-1}$$

The determination of the age of a lead mineral by any model is very dependent upon the choice of numerical values taken for the following parameters.

(a) Primeval abundances $^{206}Pb/^{204}Pb$, $^{207}Pb/^{204}Pb$, $^{208}Pb/^{204}Pb$.

(b) Age of the earth, t_0.

(c) Decay constants for ^{238}U, ^{235}U, ^{232}Th.

(d) Ratio $^{238}U/^{235}U$.

Patterson, Brown, Tilton and Inghram (1953) and Patterson, Tilton and Inghram (1955) were the first to determine the isotopic composition of lead separated from iron meteorites. Their results on troilite from the Canyon Diablo and Henbury iron meteorites are the least radiogenic leads ever observed. The isotopic compositions of primordial lead obtained by Patterson

(1955), Starik, Sobotovich, Lovtsyus, Lovtsyus and Avdzeiko (1957), Starik, Sobotovich, Lovtsyus, Lovstyus and Shats (1959), Chow and Patterson (1962) and Murthy (1961) are given in Table 46, together with the generally accepted average values. Masuda (1964) and Marshall and Feitknecht (1964) have shown that the isotopic composition of lead in iron meteorites shows significant variations. However, troilite is relatively enriched in lead and depleted in uranium and thorium; the average values $^{206}Pb/^{204}Pb = 9\cdot56$ $^{207}Pb/^{204}Pb = 10\cdot42$ and $^{208}Pb/^{204}Pb = 29\cdot71$ given by Murthy and Patterson (1962) are at present the accepted primordial lead isotope abundances. These authors quote an estimated uncertainty of $1\cdot5\%$ for the weighted values.

TABLE 46. Isotopic composition of primordial lead

Meteorites and Source	$^{206}Pb/^{204}Pb$	$^{207}Pb/^{204}Pb$	$^{208}Pb/^{204}Pb$
Patterson (1955)			
Average of Canyon Diablo and Henbury			
Chow and Patterson (1962)	9·50	10·36	29·49
Average of repeated analysis of Canyon Diablo troilite	9·61	10·39	29·87
Starik and co-workers			
Average of Canyon Diablo Burgavli, Aroos troilite	9·74	10·70	30·28
Murthy (1961)			
Sardis troilite	9·37	10·22	29·19
Average	9·56	10·42	29·71

Attempts by Gerling (1942), Holmes (1946, 1947), Houtermans (1947), Bullard and Stanley (1949), Collins et al. (1953) and Russell and Allan (1956) to calculate t_0 from the slopes of experimentally determined isochrons have not proved to be very successful.

A combination of terrestrial and meteorite leads has given more promising results. In practice, isotope ratios for minerals of known geological age are inserted in the isochron equation and solved for t_0. By this procedure, Houtermans (1953) and Patterson et al. (1955) obtained a value of $4\cdot50 \times 10^9$ years.

APPLIED GEOCHRONOLOGY

A further estimate for t_0 can be made from meteorite data alone. Lead isotope ratios in meteorites should show a linear relationship; iron meteorites will approach the primordial abundances, while stone meteorites containing significant amounts of uranium and thorium will be more radiogenic. The slope of a line joining iron and stone meteorites will be equal to their age provided that (a) the lead in stone and iron meteorites was formed at the same time, (b) the meteorites have remained closed systems with respect to Pb, U, Th since they were formed and (c) the isotopic composition of U and Pb in stone and in iron meteorites was the same at the time of their formation.

The "primary isochron of zero age for meteorites" (sometimes called a geochron) obtained by this method has a slope of $0 \cdot 59 \pm 0 \cdot 01$ corresponding to an age of $4 \cdot 55 \pm 0 \cdot 03 \times 10^9$ years. Ostic, Russell and Reynolds (1963) have obtained values of $4 \cdot 52 \pm 0 \cdot 03$ million years and $4 \cdot 55 \pm 0 \cdot 03$ million years for the age of the earth by comparing the growth of single stage terrestrial leads

TABLE 47. The age of meteorites from $^{206}Pb/^{204}Pb$ against $^{207}Pb/^{204}Pb$ ratios

Source	Number of analyses	Slope and standard deviation		Age $\times 10^9$ years	
		A	B	A	B
Patterson	5	$0 \cdot 6027 \pm 0 \cdot 0090$	$0 \cdot 5945 \pm 0 \cdot 0087$	$4 \cdot 58 \pm 0 \cdot 03$	$4 \cdot 56 \pm 0 \cdot 02$
American results	24	$0 \cdot 6045 \pm 0 \cdot 0120$	$0 \cdot 5922 \pm 0 \cdot 0095$	$4 \cdot 58 \pm 0 \cdot 03$	$4 \cdot 55 \pm 0 \cdot 02$
Russian results	28	$0 \cdot 5578 \pm 0 \cdot 0293$	$0 \cdot 5998 \pm 0 \cdot 0160$	$4 \cdot 47 \pm 0 \cdot 08$	$4 \cdot 57 \pm 0 \cdot 03$
Average		$0 \cdot 5950 \pm 0 \cdot 0117$	$0 \cdot 5944 \pm 0 \cdot 0082$	$4 \cdot 56 \pm 0 \cdot 03$	$4 \cdot 56 \pm 0 \cdot 02$

A = Linear regression analysis through observed data
B = Linear regression analysis through accepted primeval abundances
$^{206}Pb/^{204}Pb = 9 \cdot 56$, $^{207}Pb/^{204}Pb = 10 \cdot 42$
(From Ostic, 1963.)

with the growth of radiogenic lead in meteorites. The calculation did not require an estimate to be made for the ages of any of the samples used, and it did not require the assumption that the age of the earth is the same as the age of meteorites.

Kanasewich and Slawson (1964) have used meteorite data in a series of least squares analyses in order to evaluate the slope of the "primary isochron", the results of which are given in Table 47.

The numerical values used for decay constants and the ratio $^{238}U/^{235}U$ have been described in Chapter 6.

C. ANOMALOUS LEADS

In terms of dating the age of mineralization by lead isotope studies, it would be hoped that the model age would agree with the geological age. If agreement is not obtained, then either the model is incorrect or some additional factor has disturbed the "working" of the model. The models are based upon the single stage growth of lead from a particular U–Pb,Th–Pb environment with time. In Nier's original analysis, lead from Joplin, Missouri was enriched in radiogenic lead such that the model age was far too young for the geological environment in which the lead was found. On the other hand, lead from the cryolite body of Ivigtut, W. Greenland was much older than the known geological age.

If agreement with known geological age is to be used as a criterion for the validity of a particular model, then leads from Joplin and Ivigtut are anomalous in terms of the expected lead isotope abundances.

Houtermans and his co-workers recognize two types of anomalous lead.

1. B-Type (Bleiberg)—old lead occurring in younger rocks.

2. J-Type (Joplin)—young leads whose model age is too young compared with the known age of mineralization. Such leads frequently yield negative ages.

Russell and his co-workers only recognize the J-type lead as anomalous and consider that all non-conformable lead deposits are potentially anomalous. The two types of anomaly are quite different; the J-type may be considered truly anomalous as such lead has not developed in a constant U–Pb,Th–U environment according to a single stage growth curve; additional radiogenic lead has been added to the system from some other source. B-Type leads may have developed according to a single stage of growth, but represent leads that have been rejuvenated and emplaced into younger rocks without contamination by the addition of radiogenic lead during "migration".

Anomalous leads have been described by Holmes (1947), Damon (1954), Vinogradov, Tarasov and Zykov (1957), Zhirov and Zykov (1958), Houtermans and Eberhardt (1960), Russell, Kollar and Ulrych (1961), Kanasewich and Slawson (1964) and Kanasewich (1962).

Anomalous leads of the Joplin type must have been developed through at least two stages. Houtermans (1953), Geiss (1954), Russell *et al.* (1954) and Bate and Kulp (1957), Bate *et al.* (1957) have described equations relating anomalous leads, while Kanasewich (1962) and Kanasewich and Slawson (1964) have presented a model by which a quantitative determination can be made of the time of the second stage of mineralization. The general equations for the development of all leads are as follows.

$$\frac{^{206}Pb}{^{204}Pb} = \left(\frac{^{206}Pb}{^{204}Pb}\right)_{t_m} = \left(\frac{^{206}Pb}{^{204}Pb}\right)_{t_0} + 137 \cdot 8 \int_0^{t_0} \frac{^{235}U}{^{204}Pb} \, \lambda e^{\lambda t} dt$$

$$\frac{^{207}Pb}{^{204}Pb} = \left(\frac{^{207}Pb}{^{204}Pb}\right)_{t_m} = \left(\frac{^{207}Pb}{^{204}Pb}\right)_{t_0} + \int_t^{t_0} \frac{^{235}U}{^{204}Pb} \, \lambda' e^{\lambda' t} dt \qquad (33)$$

$$\frac{^{208}Pb}{^{204}Pb} = \left(\frac{^{208}Pb}{^{204}Pb}\right)_{t_m} = \left(\frac{^{208}Pb}{^{204}Pb}\right)_{t_0} + \int_t^{t_0} \frac{^{232}Th}{^{204}Pb} \, \lambda'' e^{\lambda'' t} dt$$

For primary leads, the geochemical ratios $^{235}U/^{204}Pb$, $^{232}Th/^{238}U$ have always been the same. Anomalous lead models take into account variations in these two ratios. In the Kanasewich model, the $^{235}U/^{204}Pb$ and $^{232}Th/^{238}U$, referred to by the symbols V and W respectively, are assumed to have had the values V_1 and W_1 from t_0 to t_1, V_2 and W_2 from t_1 to t_2 and V_3 and W_3 from time t_2 to t_3. With these assumptions, the equations for the growth curves can be written in the following form.

$$\left(\frac{^{206}Pb}{^{204}Pb}\right)_{t_m} = \left(\frac{^{206}Pb}{^{204}Pb}\right)_{t_0} + 137 \cdot 8 V_1(e^{\lambda t_0} - e^{\lambda t_1}) + 137 \cdot 8 V_2(e^{\lambda t_1} - e^{\lambda t_2}) +$$
$$137 \cdot 8 V_3(e^{\lambda t_2} - e^{\lambda t_3}) +$$

$$\left(\frac{^{207}Pb}{^{204}Pb}\right)_{t_m} = \left(\frac{^{207}Pb}{^{204}Pb}\right)_{t_0} + V_1(e^{\lambda' t_0} - e^{\lambda' t_1}) + V_2(e^{\lambda' t_1} - e^{\lambda' t_2}) + V_3(e^{\lambda' t_2} - e^{\lambda' t_3}) + \quad (34)$$

$$\left(\frac{^{208}Pb}{^{204}Pb}\right)_{t_m} = \left(\frac{^{208}Pb}{^{204}Pb}\right)_{t_0} + W_1(e^{\lambda'' t_0} - e^{\lambda'' t_1}) + W_2(e^{\lambda'' t_1} - e^{\lambda'' t_2}) + W_3(e^{\lambda'' t_2} - e^{\lambda'' t_3}) +$$

where $V = {}^{235}U/^{204}Pb$, $W = {}^{232}Th/^{204}Pb$, t_0 = age on the earth, t_1 = time of first mineralization, t_2 = time of second mineralization, t_3 = time of third mineralization. At any time $t_2 > t_1$, leads in the system are linearly related according to equation (34). t_1 can be considered to represent the time at which ordinary leads were differentiated from their source and either concentrated to form a lead deposit or were disseminated throughout a host rock. At the same time, uranium and thorium would enter the same environment in which the lead occurs. At t_2, a second phase of mobilization of the lead or potential lead ores occurs, and some of the ordinary lead will become contaminated with radiogenic lead formed during the period t_1–t_2 to form anomalous leads. The degree of contamination will depend upon the extent of mixing, the maximum being for the lead which is disseminated throughout a host rock. The isotope ratios of the ordinary leads which formed at time t_2 are denoted by $(^{206}\text{Pb}/^{204}\text{Pb})t_0$ and $(^{207}\text{Pb}/^{204}\text{Pb})t_0$. Equation (34) is of the same form as the equations for the growth curve of ordinary leads, and forms a straight line of slope S where

$$S = \frac{\left(\dfrac{^{207}\text{Pb}}{^{204}\text{Pb}}\right)_{t_m} - \left(\dfrac{^{207}\text{Pb}}{^{204}\text{Pb}}\right)_{t_0}}{\left(\dfrac{^{206}\text{Pb}}{^{204}\text{Pb}}\right)_{t_m} - \left(\dfrac{^{207}\text{Pb}}{^{204}\text{Pb}}\right)_{t_0}} = \frac{[(\lambda' t_1) - \exp(\lambda t_2)]}{137 \cdot 8[(\exp(\lambda t_1) - \exp(\lambda t_2)]} \tag{35}$$

Equation (35), described by Russell and Farquhar (1960a), has been extended by Kanasewich (1962).

If the leads are then extracted at t_2 they still retain this linear relationship. A plot of lead isotope ratios will give a straight line having a slope S which intercepts the primary growth curve at times t_1 and t_2. The extension of points along this straight line is due to variations in the uranium-to-lead ratio in the second stage of lead development processes. In this two stage model, t_1 represents the time when primary lead was emplaced and t_2 the time at which anomalous lead was formed. In a similar manner, the further mixing of lead at discrete periods of time may be expressed by three- or multi-stage models.

The relation between ordinary and anomalous leads has been described in a clear manner by Geiss (1954), as illustrated in Fig. 45. By plotting $^{206}\text{Pb}/^{204}\text{Pb}$ against $^{207}\text{Pb}/^{204}\text{Pb}$, a single curve will result for a particular lead environment. Leads of the same age but that have developed in source regions having different values for V will lie along a straight line called an isochron. The origin of curves (growth lines) and lines (isochrons) will be from the point of primeval lead abundances corresponding to the age of the earth. In Fig. 45, the curve from the origin to A will be the growth path for lead in environment V_1; at time t_{m_1} (point A) the lead is moved to a new environment characterized by a U/Pb ratio V_2. The new environment V_2 is assumed to contain more uranium than V_1, although the reverse can also occur. The new curve from the origin through B to G is the growth curve for lead in environment V_2; the repetition of this process described the growth of multi-stage anomalous leads. At the time t_{m_2} corresponding to G, the lead is again moved to form a lead mineral containing negligible uranium. From time t_{m_2} the lead isotopes remain unchanged, so they can be found by subtracting a vector equal and parallel to AB from G, which gives the ratios at I. The point I no longer lies along the old growth curve, so an anomalous lead has been formed. If the event t_{m_1} had not occurred, then the lead minerals formed at t_{m_2} would have isotope ratios indicated by point H. Similarly, for uranium environments V_3 and V_4 shown in Fig. 45, the resulting isotope ratios would be J and K. It is assumed that all V environment changes have occurred at the same times t_{m_1} and t_{m_2}. All the isotope ratios H, I, J, K, fall along a single straight line as, by constructing the chords AH, BG, CF and DE, all these lines are parallel and the slope represents the ratio ^{207}Pb/^{206}Pb produced in the various environments between time t_1 and time t_2.

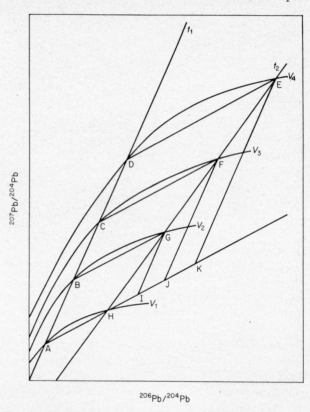

FIG. 45. Graphical illustration of the formation of anomalous leads.

This ratio must be constant over any decay time $t_{m_1}-t_{m_2}$, since this ratio for fixed values of t_1 is proportional only to the ratios ^{235}U/^{238}U in the various environments.

Chemical processes do not to any significant degree discriminate between the isotopes of uranium, and the ratio ^{207}Pb/^{206}Pb is determined solely by the two events t_{m_1} and t_{m_2}. Therefore, we have four similar parallelograms with a common angle BAH and a common direction ABCD; thus all the final points H, I, J, K, must fall on a single straight line.

In determining t_{m_1} and t_{m_2} it is necessary to consider geological and other radiogenic isotope data. When t_{m_1} can be determined, it is possible to calculate t_{m_2} according to equation (35); conversely, if t_{m_2} is known, t_{m_1} can be evaluated. The solution is best carried out by graphical analysis, although Farquhar and Russell (1963) have described a simple nomogram for finding solutions for t_1 and t_2 in equation (35). Results of calculations of anomalous leads from different areas by Kanasewich (1962) are given in Table 48. The criteria used by Kanasewich (1962) for recognizing ordinary leads for determining t_{m_1} are these.

1. The isotopic composition of a group of ordinary leads from an area should be constant within 1% or better.

2. The $^{206}Pb/^{204}Pb$ and $^{207}Pb/^{204}Pb$ ratios of an ordinary lead should have a V_0 between 0·063 and 0·067.

3. The $^{208}Pb/^{204}Pb$ ratio of an ordinary lead together with a primeval $^{208}Pb/^{204}Pb$ ratio of 24·49 should yield a $^{232}Th/^{204}Pb$ ratio between 35 and 41, calculated on the basis of a single Pb/Th system.

4. Limits set by Russell and Farquhar (1957) on the radiogenic component of an anomalous lead must not be violated. In particular, given a suite of anomalous leads for which the slope has been determined, and putting $t_2 = 0$ in equation (34), the calculated value of t_2 must be equal to or greater than the apparent age of any of the leads, ordinary or anomalous, which lie on the line. Russell, Farquhar and Hawley (1957) have pointed out that the largest possible value for t_2 is obtained when the line is a tangent to the growth curve, and the largest value of t_1 when t_2 is put equal to zero.

5. The age of an ordinary lead should agree reasonably well with other age dating methods.

Finally, the relation between lead isotopes in U–Pb environments in terms of current models may be described as follows.

$$\left(\frac{^{206}Pb}{^{204}Pb}\right)_{Today} = \left(\frac{^{206}Pb}{^{204}Pb}\right)_{Primeval} + \left(\frac{^{206}Pb}{^{204}Pb}\right)(e^{\lambda t_0} - e^{\lambda t_{m1}}) + \left(\frac{^{238}U}{^{204}Pb}\right)(e^{\lambda t_1} - e^{\lambda t_{m2}}) \ldots \quad (36)$$

$$\underbrace{}_{1} \quad \underbrace{}_{2} \quad \underbrace{}_{3}$$

Holmes, Houtermans model $= 1 + 2$
Conformable leads, Russell, Stanton, Farquhar model $= 1 + 2$
Russell, Farquhar, Cumming model $= 1 + $ average of 2
Anomalous lead $= 1 + 2 + 3 \ldots$

In recent years, much attention has been focused on lead present at trace levels in other sulphides, such as pyrite and chalcopyrite, and also that

TABLE 48. Calculated values for t_1 and t_2 on anomalous leads (Kanasewich, 1962)

Area	No. of samples	$^{206}Pb/^{204}Pb$	$^{207}Pb/^{204}Pb$	S	Standard deviation of S	t_1 (million years)	t_2 (million years)
Broken Hill	14	16·12	15·54	0·1139	0·0033	1600∓4	510∓80
Goldfield	7	15·28	15·32	0·1427	0·0063	2015∓75*	560∓250
Sudbury	11	16·03	15·61	0·1369	0·0063	1730∓100	870∓280
Ozark Dome	30	16·52	15·59	0·0886	0·0104	≈1350	≈115
New Mexico	63	16·20	15·59	0·0938	0·0029	≈1490	≈69

* Estimated error.

present in the common rock-forming silicates. The isotopic composition of lead in feldspars has been determined by Tilton *et al.* (1955), Aldrich, Wetherill, Davis and Tilton (1956), Catanzaro and Gast (1960) and Murthy and Patterson (1962), while Tilton *et al.* (1955) have determined the isotopic composition of lead in the whole rock and mineral phases of the Essonville granite, as shown in Table 49. In general, it is hoped that the isotopic composition of feldspar lead will be equivalent to that of galena lead if an ore body had formed.

When anomalous model lead ages are observed, they may be explained by (a) the presence of excess radiogenic lead on mineral surfaces. This may often be removed by mild acid leaching; (b) the mineral contains significant amounts of uranium and thorium; (c) the lead disseminated throughout a mineral is of the two stage type.

TABLE 49. Isotopic constitution of lead in the Essonville granite

	^{204}Pb	^{206}Pb	^{207}Pb	^{208}Pb
Whole rock	1·00	20·25	15·65	48·73
Whole rock, initial composition	1·00	16·4	15·4	30·2
Plagioclase	1·00	18·16	15·48	40·02
Magnetite	1·00	36·7	16·8	97·4
Apatite	1·00	31·9	16·3	76·3
Pyrite	1·00	20·34	15·92	54·34

(After Tilton *et al.*, 1955.)

Factors (a) and (b) may be assessed by measuring the present day α-particle activity. If there is no activity or if it is very small, a subsequent lead isotope analysis should approximate to the isotopic composition of the mineral lead at the time of its formation.

A further rapidly developing field is the study of trace lead in sedimentary rocks. Wampler and Kulp (1962, 1964) have obtained meaningful model ages for lead extracted from limestones, provided that the samples are enriched with respect to uranium.

Studies of lead isotopes in sediments deposited in marine environments have been made by Chow and Patterson (1959, 1962). They used pelagic sediments as a means of sampling the continental surface of the earth. In pelagic sediments, lead occurs by the direct precipitation of dissolved lead in seawater, and by mechanical transport of material denuded from the continents. If the samples that represent the average for continental crust today can be represented by pelagic sediments, they can then be used to define the average isotopic composition of lead in the crust today, by the values obtained for modern sediments are $^{206}Pb/^{204}Pb = 18.58$, $^{207}Pb/^{204}Pb = 15.7$ and $^{208}Pb/^{204}Pb = 38.87$. The results from pelagic sediments can be extended to predict the isotopic composition of lead in authigenic minerals formed in sediments. Patterson has observed a model age spread of about 400 million years for modern sediments (Tatsumoto and Patterson, 1963) and it can, to a first approximation, be assumed that this would be a feature for authigenic minerals formed in sediments throughout geological time. It is equally apparent that there will be exceptions depending upon the history of deposition in a particular basin with reference to the source of the lead. Model ages by the H.H. or R.F.C. models will tend to be less than the true age of deposition, unless the source rocks are unusually low in uranium or high in lead. The amount of radiogenic lead should be greater for near-shore sediments, or those formed in small isolated basins. On the other hand, well mixed lead in large, deep ocean basins would be expected to approximate to the crustal average at the time the sediments were deposited.

Chow and Patterson have shown that marine lead has not evolved through a single-stage growth process and the observed ratios are compatible with a two-stage development model. Using accurate stratigraphic control, it should be possible to trace variation in lead isotope abundances in the well documented stratigraphic cycles of the Time Scale.

D. TRACE ELEMENTS IN RELATION TO LEAD ORES

A mantle origin for lead ores fails to account for the association of lead with granitic, metamorphic and sedimentary rocks, rather than with basic rocks; lead is typically absent, or present in very small amounts, in primary sulphide minerals formed by the crystallization of basic magmas. Geological evidence suggests that it would appear to be more realistic to derive the lead ores from homogeneous "crustal" material. Shaw (1957) has proposed that, during the early differentiation and formation of the earth's sialic crust, uranium, thorium and lead would tend to be enriched towards the surface. In the course of time, radiogenic lead would accumulate and, during subsequent erosion, transportation and sedimentation, the ensuing sediments would contain lead with an isotope abundance characteristic of the time of deposition. This assumes that the distribution of lead in the ancient "oceanic" basins is analogous to the deposition of modern lead whose isotopic composition approximates to the present crustal average, while the local separation of U, Th and Pb would occur. Shaw suggests that, during large regional homogenization, these differences will be diluted such that large areas would act as closed systems and no permanent separation of these elements would occur. In such a case, the U/Pb,Th/U ratios would be equivalent to original crustal ratios modified only by the addition of radiogenic lead.

In the future, research into lead isotope abundances will undoubtedly show whether or not the source region of basalts contains lead of constant isotopic composition and reveal its relation to a single growth line. If oceanic leads should be characterized by excess radiogenic lead, and continental single stage leads should show features that can be related to single growth lines, then this would suggest that original differences in the isotopic composition of mantle lead have been lost during processes of crustal mixing. However, large numbers of isotope analyses will be required to settle this problem and attention should be paid to the petrological and geochemical characteristics of the samples.

In terms of the origin of lead ores there are two possibilities, a mantle and a crustal origin. The degree of enrichment of lead is another factor that must be taken into account. The question is, have some lead ores developed from primitive areas containing large amounts of lead, or are all lead ores formed by the removal of trace amounts disseminated lead in crustal or mantle rocks? An approach to this problem by studies of other major elements associated with ore deposits is difficult, as the ores are the highly fractionated end-stage of a very complex process giving rise to ore deposits, apart from other heavy metals such as Fe, Cu, Zn, Ni, Co and Ti. The common rock-forming elements such as Si, Al, Ca, Mg, N and K are absent, or present only in minor amounts. Of the abundant ore elements, zinc is of interest since most major lead deposits are in fact associated with the largest zinc deposits. A further line of approach would be a study of trace elements in ore minerals. This applies particularly to anomalous lead, in which characteristic trace elements from the environment from which the radiogenic lead was added may be retained to some degree within the lead ores. Individual trace element assemblages may not be helpful, as the ore liquid would be specific in the type of element it could retain, but absolute amounts may be more significant. For a trace element to be of use, it must occur within the crystal lattice of a galena or other common ore sulphide such as pyrite or sphalerite. A chemical analysis of a particular ore mineral may show certain elements in high abundances, but if they are present as inclusions they lose their diagnostic value, as their distribution is likely to vary considerably throughout an ore deposit. Nesterova (1958) has analysed 40 galenas from various ore deposits in the U.S.S.R. and has shown that (a) many trace elements reflect the presence of other primary minerals such as argentite, boulangerite, sphalerite, chalcopyrite and pyrite present as inclusions; (b) the following elements occur as independent minerals and do not enter the galena lattice, Ag, As, Bi, Cu, Sb and Sn; (c) recalculation of the chemical analyses shows the presence of free sulphur in most galenas irrespective of the locality; (b) trace elements in galenas and present in other sulphides appear to be characteristic for a particular region.

The problem of the chemical composition of inclusions in terms of the trace element assemblage of rocks and minerals requires to be evaluated seriously before trace elements are used in comparative studies. This can perhaps be satisfactorily approached only by use of electron probe analysis. The ultimate, which merits serious consideration, is the use of an electron microscope for defining a small area of a mineral coupled with "point source" analysis by use of a spark type of mass spectrograph.

Abnormalities in the isotopic composition of lead in sulphide minerals must be checked against the possible enrichment of U or Th in a particular phase. Vinogradov *et al.* (1958) showed that lead isotope abundances in galena, pyrite, pyrrhotite, sphalerite and chalcopyrite from one deposit were essentially the same, while cassiterite contained anomalous lead due to the presence of uranium in inclusions.

The distribution of trace elements in galenas and other sulphides has been described by Fleischer (1955), Krauskopf (1955), Cahen, Eberhardt, Geiss, Houtermans, Jedwab and Signer (1958), Mair (1958), Hawley and Nichol (1961), Doe (1962) and Ulrych and Russell (1964). Cahen *et al.* (1958) have made a systematic evaluation of trace elements in galenas related to lead isotope abundances for some N. African ore deposits. The potential value of trace elements is illustrated by relating the isotopic composition of lead to the silver content of galena. They observed that lead ores present in sedimentary rocks contained less silver than those with an obvious relation to a magmatic source and gave model ages greater than those of the host rock.

With the rapid development of lead isotope studies on lead separated from common rock-forming minerals, it is now possible to study trace element assemblages in rock leads and compare them with the trace elements in genetically related ore deposits. The isotopic composition of lead in feldspars and in co-existing galenas has been described by Tilton *et al.* (1955), Slawson and Nackowski (1959) and Catanzaro and Gast (1962).

Trace element assemblages in relation to the development of anomalous leads on a regional scale offer a fascinating field of study. By some means it is required that the common lead passes through vast thicknesses of crustal rocks removing radiogenic lead, the culmination of which is an anomalous lead deposit. The various lead isotope models can describe the beginning and culmination of this process, but they do not show how the process of lead transference occurred. It remains for geological and geochemical evidence to investigate the means and method by which this can be accomplished. If a "solution" is able to pass through a rock and extract radiogenic lead, it would also extract uranium, apart from many other elements. One would expect to find anomalous lead deposits associated with uranium deposits.

However, the lithophile character of uranium, and its chemical mobility, would probably result in the separation of the ore liquid from other elements in "solution" produced during the process of migration.

While galena is almost insoluble in an aqueous phase at room temperature, large amounts (1–600 ppm) of galena can be carried by the system $NaCl$–HCl–H_2O while, even in aqueous solutions, 20–100 ppm can be carried at 125°C. That such a system exists in ore solutions has been proved by the analysis of liquid inclusions in ore minerals that contain major amounts of sodium and chloride ions. The transport of lead in aqueous solutions is probably inadequate to account for hydrothermal ore deposits. Recent evidence suggests that hydrothermal solutions are weakly dissociated alkali-chloride electrolytes which could easily and selectively transport large amounts of lead. Such hydrothermal solutions could easily precipitate galena by reaction of the solution with country rocks during the hydrothermal alteration of the host rock. The formation of ore liquids and transport of lead have been described by Garrels (1941) and by Helgeson (1964).

IV. Case Histories

It is apparent that, while the principles of galena dating appear straightforward, their application to geological problems is often complex. For this reason, it is essential that lead isotope studies are first applied to areas that are not too complex, and in which the previous geological history is well known.

Slawson and Austin (1960, 1962) have described an area in west central New Mexico where the geology is relatively simple. Recently, lead ores derived from basement rocks have passed up vertically along a basement lineament into the overlying sediments. Lead in the sediments produced along the axis of the lineament have retained their original isotopic composition, while lead which has migrated in a lateral direction has been contaminated with radiogenic lead and is anomalous. Kanasewich (1962) has described basement lead ages of 1490 million years, while Rb–Sr, K–Ar ages by Tilton and Davis (1959) gave ages of 1300–1490 million years. Anomalous lead ages, assuming no error in the time of the first mineralization, give a Cretaceous age of $\simeq 69\pm80$ million years for the ore transference at t_2. While the error is large, this could be reduced by extending the anomalous lead line with highly radiogenic samples from the granite basement. Recent intrusive rocks of possible Tertiary age have been described by Loughlin and Koschmann (1942), and therefore the anomalous lead mineralization was either during Cretaceous or Tertiary times. This is an example in which lead isotopes can be used to define a hidden geological structure as shown in

Fig. 46. Further detailed galena studies in this area have shown that small but consistent variations in lead isotope abundances occur, indicating that less radiogenic lead was added when passing from west to east.

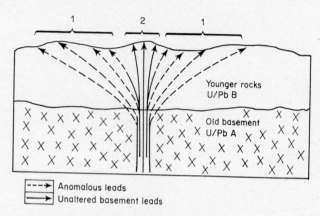

Fig. 46. Relation between isotopic composition of lead and structure. 1, Lead from U/Pb environment A that had added to it lead from U/Pb environment B. 2, Lead from U/Pb environment A, not contaminated by lead from environment B during "migration".

Cannon, Pierce and Delevaux (1963) have detected lead isotope variations in successive growth zones of a single galena crystal from the Mississippi Valley J-type Pb–Zn ore deposits. These results are illustrated in Fig. 47 and show that the radiogenic lead has been added in increments to successively younger growth layers.

The possible presence of isotopically zoned galenas requires careful studies of anomalous leads; a single grab sample may lead to inaccurate conclusions concerning two- or multi-stage lead samples.

A study of galenas from Norway by Moorbath and Vokes (1963) has shown that normal leads occur in the Pre-Cambrian basement, along the central axis of the Caledonian fold belt, and in the Permian Oslo graben. Anomalous J-type leads occur in a broad area extending from the inner part of the Caledonides east- and south-eastwards to the marginal zone of over-thrusting along the Swedish border, as well as along the margin of the Oslo province. Presumably, normal Caledonian and Permian lead has become progressively contaminated during passage through the country rocks to the site of deposition. In addition, three galenas of Houtermans B-type were also found.

A continuation of the field area into Sweden by Wickman, Blomquist, Geiger, Parwel, Ubisch and Welin (1963) has shown that the anomalous leads continue to the east into Sweden. The development of anomalous

J-type leads in such a major tectonic setting raises geochemical problems that have not been answered. A tendency often arises in moving radiogenic lead through vast thicknesses of country rock without due consideration as to the geochemical changes that must occur for many other elements. The often visualized migrating lead wave would undoubtedly remove radiogenic lead from leachable sites but, more important, corresponding amounts of

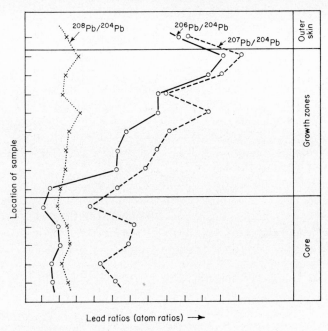

FIG. 47. Lead isotope ratios taken from points in a zoned galena crystal. (After Cannon, Pierce and Delevaux, 1963.)

uranium and thorium would also be removed. The anomalous lead deposits are not directly associated with major uranium or thorium deposits, although it is feasible that uranium may pass ahead of the lead and be deposited elsewhere. Thorium would not be expected to be as mobile as U or Pb and perhaps no major enrichment of thorium would occur. Common lead in feldspars would not be released unless complete melting occurred followed by chemical differentiation. If melting occurred, the ore body could be expected to be found in the vicinity of the remobilized rocks. An alternative explanation would be that the anomalous leads are formed by the migration of lead from below, and that the extent of radiogenic enrichment reflects local differences in the country rock. High anomalies would be equated to high U/Pb ratios for local country rocks, while less anomalous leads would reflect a lower U–Pb environment or only partial extraction of the radiogenic lead.

The main lead–zinc ores at Broken Hill, New South Wales, Australia, occur conformably in sediments that have been highly metamorphosed and then folded. Within this main ore area, the isotopic composition of galena is constant within the error of the measurement. Other ores, called the Thackeringa type, occur in an adjacent fault zone, and have distinctly different lead isotope ratios to the Broken Hill type. The associated mineral assemblage suggests the deposition of the galena at low temperatures. Russell and Farquhar (1960) have suggested that the Broken Hill type has been formed directly from the mantle; the lead isotope ratios lie on the conformable growth curve given by Stanton and Russell (1959), as shown in Fig. 44. The Thackeringa type is envisaged as being derived from the primary source by dissolution of the galena followed by its upward migration into the fault zone. Model ages for the primary mineralization give values of 1595–1604 million years by either the S.R.F. or H.H. model while the R.C.F. ages are some 200 million years lower. The age of the second mineralization, as determined by Kanasewich (1962), is found to be 510 ± 80 million years. The radiogenic contamination, as a result of the passage of the primary lead through uranium-rich country rock is reasonable, as the adjacent country rocks contain the uranium deposits of Radium Hill. Potassium–argon ages by Greenhalgh and Jeffrey (1959), Evernden and Richards (1961) and Richards and Pidgeon (1963) are between 450 and 550 million years, while rubidium–strontium ages on biotite by Richards and Pidgeon (1963) are between 460 and 550 million years. The slight re-heating that occurred approximately 500 million years ago is presumed to have provided the stimulus for the migration of the primary leads to form the secondary Thackeringa type.

Trace element studies by Fleischer (1955), have shown no difference in Ag, Mn, Fe, Si, Al, Mg, Ca content for both lead types, while Bi and Zn together with Cd, Zn, Tl, Sb were found to be present in greater amounts in the earlier conformable leads.

The complexity of lead isotope abundances is apparent when an early period of mineralization is affected by later rejuvenation. The problem becomes even more complex when rocks of many ages are involved in a major orogenic cycle. The relation between some of the worlds major Pb–Zn–Ag deposits in the Western Cordellerian chain of N. America, have been studied by, among others, Farquhar and Cumming (1954), Cannon, Stief and Stern (1958), Cannon, Pierce, Antweiler and Buck (1961, 1962), Long, Silverman and Kulp (1960), while Keer and Kulp (1952) and Eckelmann and Kulp (1957) have described uranium–lead isotope abundances from uranium mineral deposits.

The Cœur d'Alene, one of the major lead ore deposits in the northern Rockies of the U.S.A. occurs as replacement veins along shear structures that cut strongly folded sedimentary rocks of Pre-Cambrian age. Geological studies have favoured the view that the ores are of Cretacious age, and are cogenetic with the Idaho batholith. Common lead and uranium lead isotope studies indicate a Pre-Cambrian age of 1000 million years. Lead-alpha ages from small igneous intrusions present in adjacent areas give ages of about 100 million years. This pattern is consistent with the formation of the primary lead ores in Pre-Cambrian times, and the much later intrusion of granites during the Cretaceous. The six major lead deposits have model ages of between 1400 and 1270 million years while a range of 1430–100 million years is found for the smaller deposits. Originally, some of the younger leads may have had an isotopic composition identical with the major ore bodies, but have since been affected by the later addition of radiogenic lead, although it is possible that many of the intrusives may not be co-genetic.

In the Canadian Cordellera, Leech and Wanless (1962) have described the Sullivan ore body in the East Kootney District of British Columbia, the worlds largest deposit of lead and zinc. The major lead deposits occur in conformable structures, while anomalous radiogenic leads are associated with minor ore deposits and small intrusions, indicating recent remobilization of an older lead and subsequent mixing with radiogenic components. In the Sullivan ore body, lead isotope results indicate a maximum age of 1250 million years and a minimum of 765 million years, obtained by K–Ar dating of a Lamprophyre dike. Other Pre-Cambrian intrusions, such as the Hell-roaring Creek granodiorite, give a K–Ar age of 705 million years, while other granites give a Cretaceous age of 80 million years.

Although the isotopic composition of lead varies within the Cordellerian Range, the large ore bodies appear to have had a common source, and future work should be able to show whether or not a single source for the lead ore is common to both areas or whether some of the ores are unrelated and are derived from different sources.

V. The Isotopic Constitution of Sulphur in Sulphide Minerals

Sulphur is the major non-metal constituent of most ore minerals, and shows a wide range in isotopic constitution. These differences are clearly related to previous geochemical history and may shed light on an understanding of the evolution of the earth's crust and the formation of ore deposits. Investigations of sulphur isotopes were pioneered by Thode and his co-workers (Thode, Macnamara and Collins, 1949; Thode, Kleerekoper and McElcheran, 1951; Thode, Macnamara and Fleming, 1953; Thode, Wanless and Wallauch, 1954; Thode, Monster and Dunford, 1961; Smitheringale and Jensen, 1963).

Meteorite sulphur is used as a standard of comparison because of the constancy of the isotope ratio for meteorites of all types. The sulphur isotope ratio of a terrestrial sample is expressed relative to that for meteorites, as a δ-value, defined as follows:

$$\delta^{34}\mathrm{S}(\%) = \left(\frac{\left(\dfrac{^{34}\mathrm{S}}{^{32}\mathrm{S}}\right)_{\text{sample}}}{\left(\dfrac{^{34}\mathrm{S}}{^{32}\mathrm{S}}\right)_{\text{meteorite standard}}} - 1 \right) \times 10^3$$

The standard value adopted for the $^{32}\mathrm{S}/^{34}\mathrm{S}$ meteorite ratio varies between 22·20 and 22·24 (Macnamara and Thode, 1950; Vinogradov, Chupakhim and Grinenko, 1957). Similarly, the $^{32}\mathrm{S}/^{34}\mathrm{S}$ ratio for ocean water shows but little variation; and Ault and Kulp (1959) adopt a value of 21·76.

The average $^{32}\mathrm{S}/^{34}\mathrm{S}$ ratio for the earth's crust and mantle are difficult to assess, but Thode *et al.* (1961) suggest that it is close to that for meteorites (δ = 0), although Ault and Kulp (1959) estimate a value of δ = +3·6, based upon a limited number of samples. Macnamara, Fleming, Szabo and Thode (1952) considered that sulphur present in major basic rocks of deep seated origin (plateau basalts, dunites) should contain primordial sulphur. Their results indicated a slight enrichment in $^{34}\mathrm{S}$ (δ = +3%), perhaps as a result of isotope fractionation during the crystallization of the magma; on the other hand, Vinogradov (1958) has found no significant difference between meteorites and ultra basic rocks, but a similar positive enrichment of $^{34}\mathrm{S}$ in acid rocks.

Kulp, Ault and Feely (1956) have determined the $^{32}\mathrm{S}/^{34}\mathrm{S}$ ratio for many sulphide minerals of magmatic, pegmatitic and hydrothermal origin. They noticed differences between 21·53 and 23·0, and concluded that these were related to inhomogeneities in the source of the sulphur. Only sedimentary sulphides and sulphates show large deviations in $^{34}\mathrm{S}$ content compared to meteorites and basic rocks. The values for $\delta^{34}\mathrm{S}$ lie between −50 and +50% and are the result of fractionation that occurs in the oxidation–reduction processes of the biological sulphur cycle.

Shima, Gross and Thode (1963) have used the sulphur isotope method to differentiate between rocks of primary igneous origin and granitic intrusions, which are granitized or remobilized sediments. However, most granites have small, positive $\delta^{34}\mathrm{S}$ values and, at present, in these instances, it is not possible to distinguish between a crustal or mantle origin. In others, anomalously high values for $\delta^{34}\mathrm{S}$ suggest that they have been formed by remobilization of earlier sediments or by remelting of crustal material. Examples that show these differences are given in Table 50.

In the Mississippi Valley (Illinois-Iowa-Winconsin-Joplin) lead–zinc districts, no correlation exists between $^{32}\mathrm{S}/^{34}\mathrm{S}$ ratios and the $^{207}\mathrm{Pb}/^{206}\mathrm{Pb}$

ratios while, in the South-eastern Missouri district, the $^{32}S/^{34}S$ ratio increased while the $^{207}Pb/^{206}Pb$ ratio decreased, which is consistent with the assumed crustal origin of the mineralizing solutions (Kulp et al., 1956). At present, very little work has been carried out to relate and compare the ratio $^{34}S/^{32}S$ to the common lead age. Eberhardt, Houtermans and Signer (1962) have carried out a limited number of analyses on galena samples and have shown

TABLE 50. Sulphur isotope abundances in magmatic sulphide ores

Sample	$\delta^{34}S(\%)$	No. of samples	Origin
Palisades Sill, New Jersey, U.S. (Wt. mean)	+ 0·95	13	Primary
Cobalt Sill, Ontario Canada (Wt. mean)	+ 0·70	4	Primary
Leitch Sill, Ontario Canada (Wt. mean)	+ 0·10	18	Primary
Insizwa Sill, South Africa (Wt. mean)	+ 1·0	15	Primary
Dome Stack, Red Lake, Ontario, Canada	+13·30 + 6·90 +17·20 +17·20		Secondary
Rice Lake Batholith, Manitoba, Canada	+30·20 +27·20 +19·70		Secondary

(After Shima et al., 1963.)

that the two appear to be related. The continuation of such studies may lead to a means of distinguishing between different types of galena deposit and to correlate this to the problem of anomalous leads. The application of the significance of sulphur isotope ratios in relation to ore deposits may provide much useful information relating to the genesis of ore deposits. Jensen (1957, 1959) has suggested that this method may be applied to the origin of hydrothermal deposits. The sulphur ratios may provide evidence for the source

of particular ore fluids, i.e. magmatic, metamorphic, or groundwater hydrothermal deposits.

At present, a serious controversy prevails concerning the possible sedimentary or hydrothermal origin of many of the world's largest mineral
deposits. Stanton (1960) has suggested a model relating the $^{32}S/^{34}S$ ratios to
the origin of ore deposits, and has remarked that variations in the $^{32}S/^{34}S$
ratio may be effected by primary variations in the source, contamination
during migration, fractionation during formation, migration, deposition, and
homogenization, during and following deposition. Isotopic fractionation
associated with the crystallization differentiation of a tholerite magma has
been described by Smitheringale and Jensen (1957).

CHAPTER 10

DATING MODERN SEDIMENTS

I. Introduction

Dating of marine and freshwater sediments has been directed towards studying the imprint of Pleistocene changes as preserved in sediments. The foundations for studying the chronology of pelagic sediments were established between 1940 and 1950 by Piggot and Urey. It is only during recent years that modern sampling and analytical techniques have made it possible to study core samples and sediments from the ocean floor. While near-shore sediments are modern, those present in deep ocean basins may possibly penetrate into Tertiary sediments.

The age of sediments is determined by measuring the radioactivity of nuclides that were entrapped into or absorbed onto the sediments at the time of deposition. The evaluation of such ages can become meaningful only if the geochemical characteristics and the cycle of the radio nuclides are known and it is only recently that progress has been made in this field.

In the northern hemisphere, the oceans constitute 60·6%, and in the southern hemisphere 81·0% of the earth's total surface area and, apart from the Arctic Ocean (4000 ft), have a mean depth of 10–13,000 ft. Ocean sediments are wholly from terrestrial sources, apart from small, but continuous, amounts contributed by cosmic dust. Trask (1955) has studied the distribution and concentration of Sr, Ba, Cu, Ni, Pb, Mo, Re in deep sea sediments and has noted the similarity between terrestrial clays and near-shore clays, suggesting a terrigenous rather than volcanic origin. The bulk of the sediments is formed on the continental shelf and slopes, while the finer and dissolved material is carried out into the deep ocean basins where sedimentation rates are slow. Shelf sediments are occasionally carried out into the ocean by deep turbidity currents. In the deep oceans, further detritus is provided from volcanic dust, wind-blown dust, the melting of ice masses, and through the agency of man. The thickness of unaltered and undisturbed deep ocean sediments has, on seismic evidence, been estimated to be between 1000 and 1200 feet.

The oceans contain 3·2–3·6% of total salts (i.e. 32%–36% salinity index) consisting predominantly of Na, K, Mg, Ca, Sr, and SOCl, SO_4, Br, I, carbonic, boric acid, while the remaining elements of the periodic table occur in trace amounts. The simple distribution pattern of the major and trace elements dissolved in ocean water and in sediments is disturbed by the "biological cycle". This can be caused by micro-organisms removing certain elements while alive, after which they may sink to the bottom, and so enrich the sediments. In this respect, fish may extract very large amounts of some trace elements. in particular copper (0·03–0·04%), zinc (0·01%), lead (0·005–0·01%) and, in the case of the rare earths, this may amount to several percent (Arrhenius, Brambelle and Picciotto, 1957). Much of the enrichment occurs after death and provides evidence of the selective removal of trace elements present in ocean water. Apart from the absorption process, other elements such as magnesium (chlorophyll), iron (haemoglobin) and iodine (thyroxin) are essential to the metabolism of living marine organisms. In a few cases, particular organisms, such as certain holothurians and tunicates, may contain up to 10% of vanadium in their blood, an element normally present only in trace amounts. Some elements in marine samples, such as the rare-earths, show a depletion in the heavier members (Goldberg, Koide, Schmitt and Smith, 1963) relative to chondrites, while the total concentration at 4000 m shown in Table 51 is a factor of four greater than that found in surface waters.

TABLE 51. Concentration* of rare earths with ocean depth

Element	Depth	
	100 m	4000 m
Ho	$1·7 \times 10^{-4}$	$7·2 \times 10^{-4}$
Tb	$4·3 \times 10^{-4}$	28×10^{-4}
Lu	$2·0 \times 10^{-4}$	$6·4 \times 10^{-4}$

* Concentration in $\mu g/\beta$-particle. (After Goldberg et al., 1963.)

While the distribution and concentration pattern of each ocean mass may be fairly constant, each mass will have its own individual characteristics reflecting differences in the type of source material.

II. The Age of Sediments by Natural Radioactive Dating

If a radioactive nuclide enters a sediment at the time of deposition, then the specific activity of the parent or the remaining activity of a daughter product can be used to date the time of sedimentation. By dating well-defined sedimentary, horizons in different oceans, such as those of a particular warm or cold foraminiferous assemblage, or widespread glacial till, can be used to study advance and retreat stages in Pleistocene chronology.

For an isotope to be of use in dating marine sediments, it is essential that the half-life should be of the same order as that of the expected age. The most suitable radionuclides in the natural radioactive series are radium-226 ($t_{\frac{1}{2}}$ 1600 years), ionium (^{230}Th) ($t_{\frac{1}{2}}$ 80,000 years) of the ^{238}U series, protactinium-231 ($t_{\frac{1}{2}}$ 34,000 years) of the ^{235}U series, and possibly mesothorium-1 (^{238}Ra) ($t_{\frac{1}{2}}$ 6·7 years) of the ^{232}Th series.

Ocean water contains about 2×10^{-9} g/ml of uranium and 2×10^{-11} g/ml of thorium, the abundances of which are considerably lower than those found in rocks. The U/Th ratio is similarly low (Pettersson, 1937). However, the uranium (1–4 ppm) and thorium (5–15 ppm) of pelagic clays are fairly constant and similar to those found in granites. The precipitation of radium from sea-water, and its subsequent growth to the daughter product ionium, provide a method for dating sediments (Piggot and Urey, 1942). Relative to ocean-water carbonate, shells are enriched in radium (Koczy and Titze, 1958) as is fresh plankton (diatoms) in near-surface regions. In both cases, the radium may be released if the parent material should dissolve when slowly sinking in the deep oceans. This method is subject to several assumptions, foremost of which is the necessity that the radium present in the sediments at the time of deposition does not migrate from the original site of precipitation (Volchok and Kulp, 1957). Holland and Kulp (1954) have suggested an alternative process of base exchange reactions that would selectively remove radium and ionium from sea-water by the sediments. In an ideal case, the amount of radium or ionium should decrease exponentially with depth; this simple picture was not supported by Kroll (1954, 1955), who observed an irregular distribution of radium in core samples.

Bachus, Hurley and Stetson (1953) were able to show that the irregular radium concentrations were related to variation in the concentration of manganese.

A. IONIUM–THORIUM METHOD

Picciotto and Wilgain (1954) have suggested that the ionium–thorium pair would be suitable for dating, and have described a procedure for the chemical separation of ionium from ocean sediments. Results obtained by this method suggested the preferential precipitation of ionium, and that the ionium content decreases with depth, a feature not always observed from a study of radium in sediments. This is interpreted as reflecting the lack of radioactive equilibrium between ionium and radium caused by the subsequent migration of radium from the site of deposition. Baranov and Kuzmina (1957) have analysed a bottom core sample and showed that, unlike radium, the thorium content was constant, and that the amount of migration of thorium isotopes in this case at least was negligible. As long as both radioactive nuclides are precipitated in a constant proportion, the age obtained from such a ratio would be independent of the rate of sedimentation. The ionium–thorium method requires the following assumptions.

1. The ratio of $^{230}Th/^{232}Th$ in ocean water has remained constant during the age span of the sample.

2. During precipitation there has been no chemical fractionation and both ionium and thorium are present as the same chemical form.

3. The sediments should not contain any detrital material having significant amounts of either nuclide.

4. After deposition, there should be no migration of thorium in the sediments.

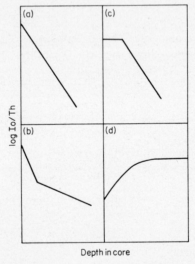

Fig. 48. Various types of Io/Th depth profiles in marine sediments. (After Goldberg and Koide, 1962.)

If all these assumptions are valid, then the ratio Io/Th should decrease with depth in the sediments. Thus ages can be calculated relative to the surface layer. In cases when the Io/Th ratio does not approach zero with depth, this indicates that uranium-supported ionium is being produced, and an appropriate correction to the Io/Th ratio is necessary.

Further errors may be related to the difference in the mean ocean residence times of uranium and thorium, which are 500,000 and 350 years, respectively.

In the latter case, this is less than the ocean mixing rate, and consequently the amount of thorium precipitated depends very much on the concentration of thorium present in unit volume of water above the accumulating sediments. Finally, the apparent precipitation rate of both ionium and thorium may be altered by biological activity in the free water, and by disturbances caused by burrowing organisms in the sediment. Goldberg and Koide (1962) have described the Io–Th method of dating and Fig. 48 illustrates how the measured profile may diverge from the ideal state.

The profiles given in Fig. 48 can be explained as follows. (a) Represents the ideal case in which there is an exponential decrease in the Io/Th ratio with depth. (b) Illustrates two discrete exponential decreases, the change occurring at a particular depth, and can be related to changes in deposition. (c) The constant values found at the surface suggest that this horizon has been disturbed by bottom water currents or through the actions of burrowing organisms. (d) A reversal of the normal decay plot suggests that at depth ionium is being produced from another source, such as uranium absorbed within the sediments and thus contained within authigenic minerals.

B. PROTACTINIUM-231–THORIUM-230 METHOD

While protactinium or thorium can be used for dating sediments, such methods encounter many difficulties and many assumptions have to be made (Koczy, 1963). The use of the ratio of protactinium to thorium should constitute a more suitable pair for dating; they are both intermediate members of the uranium radioactive series, have very similar chemical properties and hence are likely to be precipitated at the same rates under the same conditions in the sediments.

Three assumptions are required of this method.

1. The uranium ^{238}U, ^{235}U and ^{234}U should have normal abundances in sea-water.

2. The residence time for both ^{231}Pa and ^{232}Th should be comparable. The respective values are 34,300 years and 80,000 years, with a mean half-life of 60,100 years. Although the residence times are different, the ease of hydrolysis for both nuclides will reduce chemical fractionation to a minimum.

3. The uranium present in the sediments should be in equilibrium with its decay products. If uranium is present, a correction is necessary to correct for uranium-supported ^{231}Pa and ^{230}Th. The deposition of both ^{231}Pa and ^{230}Th is dependent upon the concentration of uranium in the water, and this will vary with distance from continental areas. A proportion of both nuclides will be lost in near-shore sediments, and original deposition patterns may be disturbed by the reworking of sediments through change in sea level.

If both ^{231}Pa and ^{230}Th are expressed as concentrations in uranium-units (percent equivalent or equivalent parts per million, Rosholt, 1957), and the rate of production is expressed as a ratio of the two, then the age can be calculated without knowing the uranium content. Methods for chemically separating protactinium and thorium from sea-water and sediments have been described by Rosholt (1957) and Rosholt and Dooley (1960).

The production rate for ^{231}Pa/^{230}Th is given by the equation

$$\frac{^{231}\text{Pa}}{^{230}\text{Th}} = \frac{1 - e^{-^{231}\text{Pa}t}}{1 - e^{-^{230}\text{Th}t}} \tag{37}$$

The age of a sample, corrected for uranium-supported nculides, can be obtained from equation (38).

$$t = 8 \cdot 66 \ln 2 \cdot 33 \, \frac{^{230}\text{Th}_{pv} - ^{230}\text{Th}_{s}}{^{231}\text{Pa}_{pv} - ^{231}\text{Pa}_{s}} \times 10^4 \text{ years} \tag{38}$$

where s stands for uranium supported nuclides and pv stands for uranium unsupported nuclides.

If the unsupported nuclides are determined in units of equivalent uranium, equation (38) can be simplified by subtracting the uranium content from the measured amounts of the radionuclides, as shown in equation (39).

$$t = 8 \cdot 66 \ln 2 \cdot 33 \, \frac{^{230}\text{Th} - \text{U}}{^{231}\text{Pa} - \text{U}} \times 10^4 \text{ years} \tag{39}$$

The limit of age determination by this method is about 175,000 years. Rosholt, Emiliani, Geiss, Koczy and Wangersky (1961) have determined ^{231}Pa/^{230}Th ages from two stratigraphically equivalent levels separated by a distance of 600 km; the ages obtained were in good agreement with those of carbon-14 dating. In conjunction with carbon-14 dating, the method has been used to calibrate temperature curves used in Pleistocene glaciation studies (Emiliani, 1955).

The method can also be used to study the rate of deposition of sediments, although Arrhenius (1947) and Koczy (1949) regard such measurements as rather inexact. Rates of sedimentation of a unit of sediment, expressed in units of mass per time per area, can be determined by dating the ages of the upper and lower boundaries.

C. URANIUM-234–URANIUM-238 METHOD

This method would at first appear to hold little promise, as both nuclides are chemically identical and are separated in the uranium radioactive series only by two short-lived nuclides. However, disequilibrium between ^{234}U and its parent ^{238}U has been observed by Isabaev, Usatov and Cherdyntsev (1960), Cherdyntsev, Oslov, Isabaev and Ivanov (1961) and

Thurber (1962). Theoretically this method should overcome many of the assumptions required by others; the initial $^{234}U/^{238}U$ ratio must be known and any changes in this ratio may only be as a result of radioactive decay or growth. Uranium-234 has a half-life of 248,000 years (Fleming *et al.*, 1952) and the method would be suitable over the interval 1–1·5 million years, providing a useful overlap with potassium–argon dating. The older limit is determined by experimental difficultues in measuring small variations in the $^{234}U/^{238}U$ ratio. The ratios are measured by α-pulse-height analysis, but as it is not possible to resolve the $^{234}U,^{230}Th$ α-peaks, prior chemical separation of uranium and thorium is essential. The method would be useful for Pleistocene dating of marine and lake samples, but has yet to be tested for validity.

III. The Use of Artificial Radionuclides in Marine Studies

Apart from the natural radioactivities of the uranium and thorium series, cosmic produced radionuclides are continuously being added to the earth's surface. The advent of the nuclear age has provided a further source of radioactivity through fission products produced during nuclear explosions (Miyake and Sugimura, 1961). As the oceans constitute 71% of the earth's surface, a major proportion of these radioactive nuclides will enter the ocean masses, and could be utilized for dating purposes, and also to study rates of ocean mixing. Fission products are released into the upper atmosphere, while man is continuously adding into the sea, radioactive nuclides in the form of waste products from nuclear research establishments.

Among the many nuclides produced through cosmic and man-controlled nuclear reactions, the most useful appear to be ^3H (tritium, $t_{\frac{1}{2}}$ 12·262 years; Libby, 1953; Jones, 1955; Giletti, Bazan and Kulp, 1958; Begemann, 1960; Craig, 1961), ^{10}Be ($t_{\frac{1}{2}}$ 2·5 × 10^6 years; Peters, 1955, 1957; Arnold, 1956), ^{14}C ($t_{\frac{1}{2}}$ 5·67 years; Chapter 3), ^{26}Al (6 $\pm$ 3 × 10^5 years Kohman, 1956; Ehman, 1957), ^{36}Cl (3·08 × 10^5 years; Davis and Schaeffer, 1955; Winsberg, 1956) and ^{32}Si ($t_{\frac{1}{2}}$ 710 years; Lindner, 1953; Lal *et al.*, 1960). For studies of surface-water mixing, short lived radionuclides are required while, in deep oceans, those having half-lives greater than the mean ocean residence time are necessary. For cosmic produced nuclides, it is essential that the cosmic flux has been constant over the period to be dated; Lal *et al.* (1960) has shown that the production rate of ^3H, ^{10}Be, ^{24}C, ^{32}Si has remained constant over the last few million years.

While advances in physical and chemical techniques now make it possible to detect or measure these radionuclides, further work is required to understand their geochemical behaviour in a marine environment.

CHAPTER 11

METEORITE AGES

I. Introduction

Meteorites are the only source of extra-terrestrial material, and are regarded as representing hidden layers of the earth's crust that are not available for laboratory study. The isotopic constitution of many elements in terrestrial and extra-terrestrial samples are identical and in cases where differences are observed they may be related to one of the following causes.

1. As a result of physical or chemical fractionation between isotopes; this is generally restricted to the light elements with a mass less than 40.

2. Isotopic variations directly related to radioactive decay, or the fission of long-lived isotopes.

3. As meteorites pass through outer space, they are continuously bombarded by a flux of high energy cosmic ray particles. The impacts of these particles on meteorites give rise to light and heavy particles described as spallation products.

The measurement and significance of meteorite ages have been described by Eberhardt and Geiss (1960a), Reynolds (1960a), Anders (1962), Fish and Goles (1962), Krummenacher, Merrihue, Pepin and Reynolds (1962), Schaeffer (1962), Hintenberger (1962) and Zähringer (1964).

II. Solidification Ages

The age of a primary meteorite body cannot be dated as after its formation it underwent chemical differentiation to give rise to metal and silicate phases.

However, subsequent to this event, once this stage was completed and the meteorite body solidified and became a closed system, the principles of conventional dating may be applied. Meteorite ages reflect chemical fractionation between parent and daughter elements. Chemical fractionation in meteorites is not comparable to that observed in terrestrial material, in fact in some cases the parent nuclide is depleted while the daughter product may be relatively enriched.

A. RUBIDIUM–STRONTIUM METHOD

The lithophile character of rubidium and strontium result in a preferred concentration of those elements in stone meteorites, while they are present at very low abundances in irons. Meteorites may be depleted in rubidium by vapourization, also a proportion is readily leachable (Gast, 1960a, b; Smales, Hughes, Mapper, McInnes and Webster, 1964). The $^{87}Sr/^{86}Sr$ ratio in chondrites is variable while that in calcium-rich achondrites is rather constant at about 0·700. Iron meteorites present analytical difficulties in view of the very small amounts of strontium and rubidium. The enrichment in ^{87}Sr in the chondrites is brought about by the decay of ^{87}Rb with time, while the achondrites contain such small amounts of rubidium that the ratio $^{87}Sr/^{86}Sr$ must approximate to the primordial value. If all meteorites have the same age and initial $^{87}Sr/^{86}Sr$ ratio, then a plot of $^{87}Sr/^{86}Sr$ against $^{87}Rb/^{86}Sr$ should present a series of points lying along a straight line; this isochron will give the age, and also the initial $^{87}Sr/^{86}Sr$ ratio. Depending upon the decay constant used, and whether or not the observed ratios are normalized to 0·1194 or not, the age of a meteorite suggested by the Rb/Sr method is about $4·55 \pm 0·02 \times 10^9$ years for an initial $^{87}Sr/^{86}Sr$ ratio of 0·698 (Gast, 1962; Herzog and Pinson, 1956; Pinson, Schnelgler and Beiser, 1962). If the Earth and meteorites have been formed from the same material and by similar processes, then the initial $^{87}Sr/^{86}Sr$ ratio for terrestrial material should be identical to that found in meteorites.

B. RHENIUM–OSMIUM METHOD

The rhenium–osmium method should be comparable to that of the Rb–Sr method but, apart from some uncertainty in the decay constant, their geochemistry is not well known. An isochron analysis of $^{187}Os/^{186}Os$ against $^{187}Re/^{186}Os$ gives an age of $4·0 \pm 0·8 \times 10^9$ years for an initial $^{187}Os/^{186}Os$ ratio of 0·834. The finding of very low $^{187}Os/^{186}Os$ in terrestrial samples may result in terrestrial ages older than those of the oldest known rocks.

As osmium is an exceptionally stable element, the $^{187}Os/^{186}Os$ ratio should not be affected by repeated cycles of magmatic differentiation.

9*

C. Lead Method

Lead isotope studies have two advantages over the Rb–Sr or Os–Re methods, as only isotope ratios are measured and the decay constants are known with more certainty. Patterson (1956) has been foremost in applying the method to meteorite studies and, provided that the following assumptions are valid, the method yields ages of $4 \cdot 4$–$4 \cdot 6 \times 10^9$ years.

These assumptions are (a) that all meteorites were formed at the same time, (b) that the original lead was of the same isotopic constitution, (c) that the isotopic constitution of terrestrial and meteorite uranium was identical and (d) that meteorites developed in closed and isolated systems. The analysis is similar to that of Rb–Sr dating, and pairs of meteorites are used with differing U/Pb ratios. All common lead will contain varying amounts of radiogenic lead formed by the decay of uranium and thorium; lead separated from practically uranium-thorium-free troilite is virtually primordial. If the content of both uranium and lead and also the isotopic constitution can be determined with sufficient accuracy, then ages can be obtained by using uranium–lead pairs. Uranium and lead are related by the following equation.

$$\begin{aligned}
^{207}\text{Pb}_{\text{radiogenic}} &= {}^{235}\text{U}(e^{\lambda t} - 1) = {}^{207}\text{Pb}_{\text{measured}} - {}^{207}\text{Pb}_{\text{primeval}} \\
^{206}\text{Pb}_{\text{radiogenic}} &= {}^{238}\text{U}(e^{\lambda t} - 1) = {}^{206}\text{Pb}_{\text{measured}} - {}^{206}\text{Pb}_{\text{primeval}}
\end{aligned} \qquad (40)$$

Relative to common lead dating, the method has the disadvantage that absolute quantities are measured rather than ratios.

While many advances have been made in lead dating, the results often show many anomalies, e.g. when accurate U–Th analyses are available, the measured amount of radiogenic lead is too great, also the isotopic constitution of lead may vary in different facies of the same sample.

Starik, Sobotovich, Lovtsyus, Shats and Lovtsyus (1960 have suggested that iron meteorites can be divided in two groups according to the constitution of primordial lead.

When present, the excess of radiogenic lead can be interpreted as indicating that either the iron meteorite is much older than the current estimate of $4 \cdot 50 \times 10^9$ years, or that at some time it has been depleted in uranium or had radiogenic lead added from another source. While either of these may occur in an extra-terrestrial environment, they could also be the result of terrestrial contamination, particularly that caused by weathering.

III. Gas Retention Ages

Gas retention ages can date only the time at which temperatures were low enough to allow a gas phase to accumulate. This may coincide with a primary event, although later heating, such as caused by collision, will result in the loss of gas.

A. POTASSIUM–ARGON METHOD

Gerling and Pavlova (1951) were the first to determine the age of meteorites by the K–Ar method.

The very low abundance of potassium and argon presents severe analytical problems. The ages obtained by this method cover a considerable range of about 0.7–4.6×10^9 years for both stone and iron meteorites, with a noticeable concentration of these values between 4.0 and 4.5×10^9 years.

It is probable that most meteorites crystallized at least 4.5×10^9 years ago, and that the low ages are the result of argon loss.

One of the major problems inherent to the method is the amount of radiogenic ^{40}Ar and non-cosmogenic ^{40}K present. Potassium-40 is formed as a spallation product and, in the case of some iron meteorites, Voshage and Hintenberger (1960) found the abundances of ^{39}K, ^{40}K and ^{41}K to be approximately 2:1:2, which are markedly different from those of normal terrestrial potassium ($10:10^{-3}:1$). In these meteorites, the amount of normal potassium is very small.

Stoenner and Zähringer (1958b) have determined K–Ar ages in iron meteorite by activation analysis, and have obtained ages ranging from 5 to 13 million years. Stoenner and Zähringer were able to show a correlation between ^{40}Ar and the total potassium content of the samples. This points to a genetic relationship, but the neutron activation method allows only a measurement of the ^{40}Ar content and not those of other argon isotopes. Contamination by atmospheric argon is excluded by the method used, but the possibility of the addition of primeval argon entering the stone meteorite phase during the iron–stone separation must be considered. These results were at first met with some scepticism, but further work has tended to support the earlier results and, as yet, no alternative solution has been found to account for the old ages. The samples used by Zähringer were taken from the centre of large meteorite samples, so as to reduce the contribution of ^{40}K from cosmogenic processes.

B. URANIUM–HELIUM METHOD

The uranium–helium method suffers from the same restrictions as those of the K–Ar method, namely helium loss by diffusion and helium formed by cosmogenic processes. The uranium and thorium content of iron meteorites is too low, apart from troilite phases, to be of much use, but the method is practical for the chondrites. The ages show a spread similar to those found by the K–Ar method and, apart from irons, calcium-poor achondrites and pallasites, where cosmogenic ^4He may exceed 50%, the U–He ages tend to be lower than those obtained by K–Ar dating, suggesting a greater ease of helium diffusion.

IV. Nuclear Reactions by Cosmic Radiation

The impact of cosmic radiation on meteorites during flight results in nuclear transformations and solid state effects (thermoluminescence). The former are produced by high energy primary radiation as well as secondary particles, such as mesons and neutrons produced within the meteorite; the results are called spallation products. Spallation product studies provide information relating to the intensity of bombardment, spatial and temporal constancy of cosmic radiation, shape and size of a meteorite before entering the earth's atmosphere and the length of time that has elapsed since the meteorite fell. A list of cosmic ray produced isotopes detected in meteorites is given in Table 52 (Eberhardt and Geiss, 1960b).

TABLE 52. Isotopes produced by cosmic ray impact on meteorites

Isotope	Half-life	Isotope	Half-life
^{3}He		^{45}Ca	164·0 days
^{4}He		^{57}Co	270·0 days
^{20}Ne		^{54}Mn	291·0 days
^{21}Ne		^{49}V	330·0 days
^{22}Ne	Stable	^{22}Na	2·58 years
^{36}Ar		^{60}Co	5·24 years
^{38}Ar		^{3}H	12·30 years
^{41}K		^{44}Ti	∼200·00 years
^{45}Sc		^{39}Ar	325·00 years
^{48}V	16·0 days	^{32}Si	∼700·00 years
^{51}Cr	27·8 days	^{59}Ni	8×10^4 years
^{37}Ar	35·0 days	^{36}Cl	$3·1 \times 10^5$ years
^{58}Co	71·0 days	^{26}Al	$7·4 \times 10^5$ years
^{56}Co	77·0 days	^{53}Mn	$\geqslant 2 \times 10^6$ years
^{46}Sc	84·0 days	^{10}Re	$2·5 \times 10^6$ years
		^{49}K	$1·25 \times 10^9$ years

Comparative studies of different radioactive spallation products with different half-lives have shown that the cosmic flux has remained constant over the last several million years. As the cosmic radiation will only penetrate to a certain depth into the meteorite, the amount of spallation products will decrease with depth. Fireman (1958) and Hoffman and Nier (1958, 1959) determined the distribution of some spallation products in cross-sections of meteorites. By contouring areas of equal concentration they were able to illustrate the pre-atmospheric shape of the meteorites, and estimate the amount that had been ablated during passage through the earth's atmosphere.

The distribution of ^{3}He in the Grant meteorite is given in Fig. 49 (Fireman, 1958).

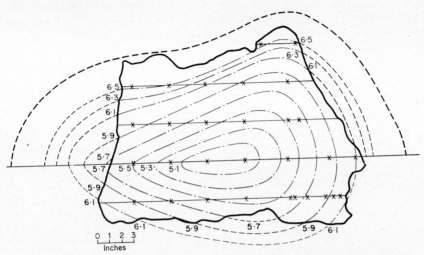

FIG. 49. Contours of constant ^{3}He content in Grant meteorite. (After Fireman, 1958.)

A radiation age can be calculated by use of a pair of isotopes, one radioactive and the other stable, according to the equation:

$$T = \frac{1}{\lambda} \frac{P_R}{P_S} \frac{S}{R}$$

(Eberhardt and Geiss, 1960b), where R and S are the number of atoms of the radioactive and stable products respectively, P_S and P_R are the effective macroscopic production cross-sections and λ the decay constant of the radioactive isotope. The pairs ^{3}H–^{3}He, ^{22}Na–^{22}Ne, ^{26}Al–^{36}Cl and ^{10}Be–^{26}Al have been widely used in measuring radiation ages.

Various analytical procedures are used to measure the abundances of spallation products: e.g. K, Ca, V, rare gases (^{3}H, Ne, Xe) are determined by mass spectrometry, Sc by radioactivation analysis, and Cl, Al, C by low level counting techniques.

The production of cosmogenic radioactivity stops after the meteorite passes through the earth's atmosphere and, after impact, the induced radioactivities decay according to their respective half-lives. By comparing the activity of a particular radionuclide from a meteorite whose fall date is known to one that is unknown, it is possible to date or place a lower limit on the fall age of the latter.

Eberhardt and Geiss (1963) have established a relation between classes of meteorites and radiation ages for chondrites. All radiation ages are very much lower than those obtained by radiometric methods, and indicate that individual meteorites have spent most of their existence within a larger body and, consequently, have been shielded from cosmic radiation. The radiation ages for iron meteorites are generally higher than those for stones; the octahedrites give ages of about 500 million years and the hexahedrites about 300 million years (Anders, 1962). However, no clear-cut age distribution has been observed between radiation ages for stones and irons, and both show a continuous and extended distribution pattern (Kirsten, Krankowsky and Zähringer, 1963). Radiation age differences have been observed by Eberhardt and Geiss (1963) between the two Urey-Craig (1953) high-iron and low-iron groups of chondrites. The low-iron group was formed 20–25 million years ago in a collision involving at least 65 km³ (Eberhardt and Geiss, 1963) of low-iron group material. The average life span of a stony meteorite is about 15 million years, hence the present abundance of meteorite classes approximates to an average of the material involved in larger collisions during the last 30 million years.

V. Extinct Radioactive Nuclides in Meteorites

During the process of nucleosynthesis, radioactive isotopes were undoubtedly formed and, if they had short half-lifes and have since decayed, their presence should be detected by abnormalities in the isotopic constitution of the stable nuclides. Brown (1947) was the first to suggest this approach of extending dating into the period of nucleosynthesis.

Suess (1948) first discussed the possible significance of radioactive ^{129}I with a half-life of 17 million years, produced during the principal galactic synthesis and which decays to ^{129}Xe. Reynolds (1960) was the first to show that excess xenon was present in the Richardson chondrite, and that the meteorites Richardton and Abee began to retain xenon 52 million years after the cessation of nucleosynthesis. The Bruderheim meteorite was formed 17 million years earlier, while troilite from the iron meteorite Sardis postdates Abee by 200 million years (Reynolds, 1963). The mass spectrum of xenon extracted from the Richardton stone meteorite is given in Fig. 50; the horizontal lines show the spectrum for terrestrial xenon (Reynolds, Merrihue and Pepin, 1962). Dating by means of excess xenon-129 (xenology) assumes that the excess xenon was formed within the meteorite at the time of cooling. Studies of other rare gases show that primeval gas is occluded in some meteorites and the excess xenon may in fact be primeval gas from an atmosphere in which ^{129}I decay has contributed more ^{129}Xe than it has in the terrestrial atmosphere. Eberhardt and Geiss (1960a) have suggested that, if this is true, then xenon dating can have no cosmological significance. Jeffrey and Reynolds (1961) have been able to show that in Abee the excess xenon was produced

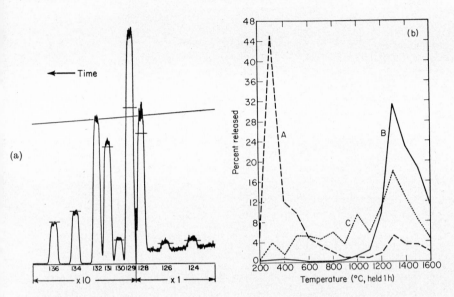

FIG. 50. (a) Mass spectrum of xenon extracted from Richardton stone meteorite (after Reynolds, 1960). (b) Release of xenon isotopes from neutron-irradiated meteorite Richardton with heating. A, Excess ^{128}Xe from (nγ) on ^{127}I. B, Excess ^{131}Xe from (nγ) on ^{130}Te. C, Excess ^{129}Xe from decay of extinct ^{129}I. (After Reynolds, 1963.)

in situ, while xenon abundances in Richardton also show less striking departures from normality as shown in Fig. 50(b). Other extinct nuclides such as ^{205}Pb and ^{244}Pu might become important in the chronology of the early history of the planetary system (Clarke and Thode, 1963).

CHAPTER 12

GEOLOGICAL TIME SCALE

The geological time scale is generally taken to refer to post-Cambrian rocks (Kulp, 1960), in which it is possible to relate relative ages obtained from the progressive development of organic life to ages obtained by radiometric methods. The Cambrian period represents the earliest time from which the gradual evolution of fauna and flora can be traced. However, organic algal remains have been found in early Pre-Cambrian sediments, although they do not exhibit changes that can be used in time scale studies. Organic remains have been studied in great detail, and for a species to be useful for dating it should have existed over wide areas, during long periods of time, and show gradual morphological variations with time. These variations are of importance only when they can be related over reasonable periods of time, and they must be clearly distinguished from local or short-term variations as a result of sudden changes in the local environments.

In many extensive basement granite areas, relative age patterns can be obtained by reference to intrusive events and structural fold patterns, but it is rare for such field evidence to enable the age of a particular event to be dated with any certainty. Often, broad estimates of relative age patterns can be deduced in Pre-Cambrian terrains by reference to some major geological event. An example of this would be the recognition over large areas of an early Pre-Cambrian sequence of sediments, cut and metamorphosed by later granites, themselves overlain by a different sequence of sediments which in turn are metamorphosed and overlain by fossiliferous Cambrian sediments. In such a case, it may be possible to divide Pre-Cambrian time into Early, Middle and Late periods, but it is not possible to give a precise age to any member. Estimates can be made on rates of deposition of the various components of sediments, but the presence of major unconformities can lead to false ages. Alternatively, the Pre-Cambrian can be subdivided in some areas by reference to a marker horizon, such as a tillite bed or major widespread extrusion of basic rocks.

The application of radiometric dating methods has now made it possible to date most igneous and metamorphic rocks. However, the measured ages must be interpreted correctly if they are to be of use in time scale studies. Much caution must be applied in assessing the significance of a single radiometric age, and it is only when several or many ages are measured that they can be regarded as significant. With improvements in analytical methods, it is now possible to date Recent rocks and minerals. Much attention has recently been focused on dating post-Tertiary samples, and in this field the K–Ar method has made outstanding advances. Damon and Bikerman (1964) have shown that K–Ar ages can be used to date a series of young volcanic extrusions which do not contain intercalated fossiliferous horizons. They have shown that a sequence of intrusive and extrusive events in the Tucson Mountains, Arizona, U.S.A. were only partly correct. In a profile through the Southern Tucson Mountains, illustrated in Fig. 51, the Cat Mountain

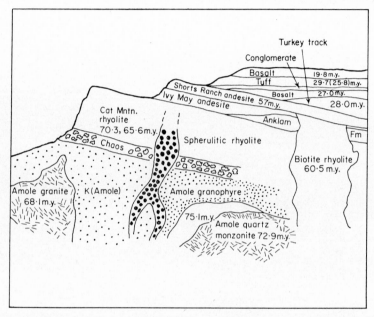

FIG. 51. K–Ar ages of S. Tucson Mts., Pima Co., Arizona, U.S.A. (After Damon and Bikerman, 1964.)

rhyolite is established as Upper Cretaceous in age rather than Tertiary, and the igneous activity is confined to two separate periods, the Laramide interval and the Middle Tertiary, Upper Oligocene Miocene interval, rather than to a single continuous event.

In 1947, Holmes was the first to construct a time scale based upon the complete isotopic determination of five samples that were analysed by Nier. Holmes (1960) and Kulp (1960) have made a preliminary revision of the time scale based upon material available in 1959. In recent years, many new ages have been measured that place close limits upon various divisions and sub-divisions of the time scale, but for present purposes that given by Kulp (1961) in Fig. 52 will suffice.

Seven-eighths of the earth's history occurred before the Cambrian, and consists mainly of granite and metamorphic rocks. It is only through the use of radiometric dating methods that it has been possible to determine the

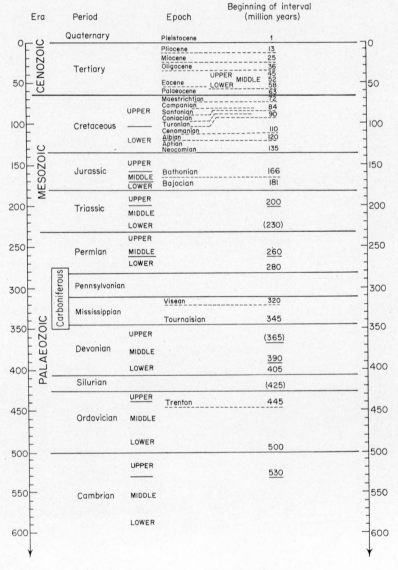

Fig. 52. Geological time scale. (After Kulp, 1961.)

age of these rocks. While major advances have been made in dating Pre-Cambrian rocks, progress has been slow in measuring the age of post-Cambrian rocks. This has been caused by the lack of suitable material or geological settings.

The direct dating of sedimentary rocks with diagnostic flora or fauna is complicated by the presence of detrital minerals that contain significant amounts of daughter nuclides used in the major methods of geochronology. If the daughter nuclide is a gas, then it is often easily lost by diffusion or through the general processes of weathering. It is only in the case of a true authigenic mineral that some certainty can be placed on the measured age. Glauconite has proved to be the most suitable mineral, and has been widely used by Russian workers to date sediments from the Recent to early Pre-Cambrian. Another approach has been to use potassium-rich volcanic rocks or ash beds intercalated between fossiliferous sediments.

In the fossil-bearing post-Cambrian rocks, the age of sediments may be determined by correlation with igneous intrusive rocks which cut the sediments. For precise dating, the interval between the intrusive event and the subsequent deposition of sediments, should be small. In an example of this approach, Faul (1960) has obtained a useful time scale point for the Vosges granite (France). The granites intrude and metamorphose Tournaisian sediments, and are overlain by fossiliferous strata of early Visean age. The Rb–Sr age of 322 ± 5 million years and K–Ar age of 315 ± 5 million years closely date the Visean–Tournaisian boundary.

A Pre-Cambrian time scale (covering the period from 600 to ~ 3500 million years) similar to that developed for post-Cambrian rocks does not exist. At present, it is possible to record profound events manifested on a world-wide scale. Vinogradov and Tugarinov (1961) have suggested a scheme for the sub-division of the Pre-Cambrian, while Wasserburg (1961) has described some of the problems associated with Pre-Cambrian geochronology.

The publication of a large number of age measurements now makes it possible to examine age patterns with time in Pre-Cambrian terrains. The results only become meaningful when they can be supported by detailed field relations and structural studies.

The North American continent can be divided into well defined age provinces and although, within each province, significantly younger or older ages may be found, they are in the minority. This pattern shown in Fig. 53(a) (Gastil, 1960) illustrates a general decrease in the age of rocks from the central areas to the present margin. Wilson (1949) has suggested that this illustrates a process of continental accretion. However, the presence of old age at the margin of the present continent does not support this view, but rather one of a gradual retreat of crustal instability from the interior outwards.

Some outstanding papers illustrating age patterns for the Pre-Cambrian
of Russia and adjacent areas have been described by Gerling and Polkanov
(1958), Komlev (1958) and Vinogradov, Tarasov and Zykov (1959).

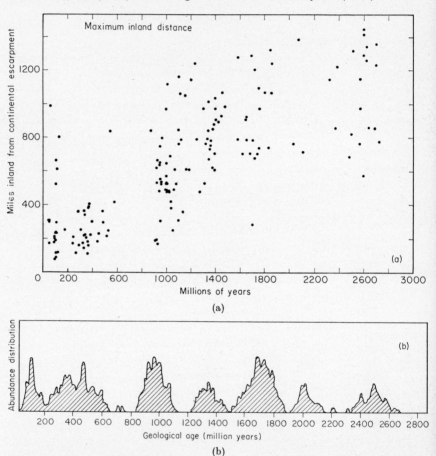

FIG. 53. (a) Variation of geological age with distance from continental escarpment,
N. America. (b) Episodic nature of mineral ages with geological time. (After Gastil, 1960.)

The distribution of mineral ages throughout geological time described by
Gastil (1960) and illustrated in Fig. 53(b) show that the major periods of
orogeny and igneous activity occur at well defined periods in all continents
throughout geological time. When large structural areas are considered, they
can often be defined by a narrow age range, indicating the last time at which
the area experienced crustal activity.

It should now be possible to produce a universal Pre-Cambrian time scale
based upon orogenic periods, but the tendency to describe these orogenies

in terms of local nomenclature leads to a confusing array of names depicting a universal event.

The post-Cambrian time scale can now be used to correlate age problems in palaeontology, orogenic movements, and periods of mineralization. While many minor differences still exist in limiting the extent of Periods and Epochs, they are slowly being resolved. While the Pre-Cambrian time scale is not known in such detail as that of post-Cambrian times, it is possible to recognize major orogenic events and to correlate them with periods of mineralization. With current advances in radiometric dating, it would be more practical to extend the restricted range of the time scale back into the Pre-Cambrian.

Radiometric dating has been able, in a few areas, to probe back to about 3500 million years, but the age of the earth is at least 4500 million years, leaving approximately 1000 million years unaccounted for. This is probably a reflection of the instability of the crust, a region that is continuously being deformed. A time is reached in any primitive crust when the original ages are permanently obliterated by the crust being repeatedly deformed by later events.

REFERENCES

Abbey, S. and Maxwell, J. A. (1960). *Chemistry in Canada* **12**, No. 9, 37.

Adams, J. A. S. and Maeck, W. L. (1954). *Analyt. Chem.* **26**, 1635.

Adams, J. A. S., Osmond, J. P. and Rogers, J. J. W. (1959). *Phys. Chem. Earth* **3**, 298.

Adams, J. A. S., Richardson, J. E. and Templeton, C. C. (1958). *Geochim. cosmochim. Acta* **13**, 270.

Ahrens, L. (1946). In "Rept. of the Committee on the Measurement of Geologic Time (1946–1947)", ed. by J. P. Marble, p. 47.

Ahrens, L. H. (1948). *Phys. Rev.* **74**, 74.

Ahrens, L. H. (1951). *Geochim. cosmochim. Acta* **1**, 312.

Ahrens, L. H. (1952). *Appl. Spectrosc.* **6**, No. 5, 11.

Ahrens, L. H. (1955a). *Geochim. cosmochim. Acta* **7**, 294.

Ahrens, L. H. (1955b). *Geochim. cosmochim. Acta* **8**, 1.

Ahrens, L. H. (1956). *Phys. Soc. Rep. Progr. Phys.* XIX, 80.

Ahrens, L. H. and Taylor, S. R. (1963). "Spectrochemical Analysis", 2nd Ed. Addison Wesley Press, Cambridge, Mass.

Akishin, P. A., Nikitin, O. T. and Panchenkov, G. M. (1957). *Geokhimiya*, No. 5, 500.

Aldrich, L. T. and Nier, A. O. (1948). *Phys. Rev.* **74**, 876.

Aldrich, L. T. and Wetherill, G. W. (1958). *Ann. Rev. nucl. Sci.* **8**, 257.

Aldrich, L. T., Herzog, L. F., Abelson, P. H. and Bolton, E. T. (1952). *Phys. Rev.* **87**, 186.

Aldrich, L. T., Wetherill, G. W., Davis, G. L. and Tilton, G. R. (1956). *Bull. Amer. phys. Soc.* Ser. 11, **1**, 31.

Aldrich, L. T., Wetherill, G. W., Davis, G. L. and Tilton, G. R. (1958). *Trans. Amer. geophys. Un.* **39**, 1124.

Aldrich, L. T., Herzog, L. F., Holyk, W. K., Whiting, F. B. and Ahrens, L. H. (1953). *Phys. Rev.* **89**, 631.

Allan, D. W., Farquhar, R. M. and Russell, R. D. (1953). *Science* **118**, 486.

Allen, J. S. (1939). *Phys. Rev.* **11**, 966.

Allsop, H. L. (1961). *J. geophys. Res.* **66**, 1499.

Alpert, D. (1953). *J. appl. Phys.* **24**, 860.

Alpher, R. A. and Herman, R. C. (1951). *Phys. Rev.* **84**, 1111.

Amirkhanoff, K., Brandt, S. B. and Bartnitsky, E. N. (1961). *Ann. N.Y. Acad. Sci.* **91**, Art. 2, 235.

Anders, E. (1962). *Rev. mod. Phys.* **34**, No. 2, 287.

Anderson, E. C. and Libby, W. F. (1951). *Phys. Rev.* **81**, 64.

Anderson, E. C., Libby, W. F., Weinhouse, S., Reid, A. F., Kirshenbaum, A. D. and Grosse, A. V. (1947). *Science* **105**, 576.

Arnold, J. R. (1956). *Science* **124**, 584.

Arnold, J. R. and Anderson, E. C. (1957). *Tellus* No. 1, **9**, 28.

Arnold, J. R. and Libby, W. F. (1949). *Science* **110**, 678.

Arrhenius, G. (1947). *Bull. geol. Soc. Amer.* **65**, 1228.

Arrhenius, G., Brambelle, M. N. and Picciotto, E. (1957). *Nature* **180**, 85.

Aston, F. W. (1927). *Nature* **120**, 224.

Aston, F. W. (1942). "Mass Spectra and Isotopes", 2nd Ed. Edward Arnold, London.

Ault, W. U. and Kulp, J. L. (1959). *Geochim. cosmochim. Acta* **16**, 201.

Baadsgaard, H., Lipson, J. and Folinsbee, R. E. (1961). *Geochim. cosmochim. Acta* **25**, 147.

230   APPLIED GEOCHRONOLOGY

Bachus, M. M., Hurley, P. M. and Stetson, H. C. (1953). *Bull. geol. Soc. Amer.* **64**, 1391A.
Badalor, S. T., Basitova, S. M. and Godunova, L. I. (1962). *Geokhimiya*, No. 9, 934.
Bähnisch, J. (1955). *Z. Phys.* **142**, 565.
Bainbridge, K. T., and Nier, A. O. (1950). Prelim. Rep. No. 9, Nucl. Sci. Ser. Nat. Res. Coun. U.S., Washington, D.C.
Ballaria, C. (1955). *Science* **121**, 409.
Baranov, V. I. and Kuzmina, L. A. (1957). Proc. 1st UNESCO Int. Conf. Sci. Res., ed. by Extermann, Vol. 2, p. 619.
Barendsen, G. W. (1957). *Rev. sci. Instrum.* **28**, 430.
Barnard, G. P. (1953). "Modern Mass Spectrometry". The Institute of Physics, London.
Barnard, G. P. (1956). "Mass Spectrometer Researches". H.M.S.O., London.
Barrell, J. (1917). *Bull. geol. Soc. Amer.* **28**, 745.
Bate, G. L. and Kulp, J. L. (1957). Lamont Geol. Obs. Tech. Rep. Cont. AT(30-1)-1114.
Bate, G. L., Miller, D. S. and Kulp, J. L. (1951). *Analyt. Chem.* **29**, 84.
Bate, G. L., Gast, P. W., Kulp, J. L. and Miller, D. S. (1957). Manuscript, Lamont Observatory, N.Y.
Bauer, C. A. (1947). *Phys. Rev.* **72**, 354.
Beard, G. B. and Kelly, W. H. (1961). *Nucl. Phys.* **28**, 570.
Begemann, F. V. (1960). "Nuclear Geology", Varenna Conf., p. 109.
Begemann, F. V., Buttlar, H., Houtermans, F. G., Isaac, N. and Picciotto, E. (1953). *Geochim. Cosmochim. Acta* **4**, 21.
Belcher, R. and Wilson, C. L. (1955). "New Methods in Analytical Chemistry". Chapman Hall.
Benson, S. W. (1960). "The Foundations of Chemical Kinetics". McGraw-Hill, New York.
Bentley, P. G., Bishop, J., Davidson, D. F. and Evans, P. B. F. (1959). *J. sci. Instrum.* **36**, 32.
Beukelman, T. E. and Lord, S. S. (1960). *Appl. Spectrosc.* **14**, 12.
Beynon, J. H. (1960). "Mass Spectrometry and its Use in Organic Chemistry". Elsevier, Amsterdam.
Bien, G. S., Rakestraw, N. W. and Suess, H. E. (1962). I.A.E.A. Symposium, Athens, Greece.
Birch, F. (1951). *J. geophys. Res.* **56**, 107.
Bishop, J., Davidson, D. F., Evans, P. B. F., Hamer, A. N., McKnight, J. A. and Robbins, E. J. (1960). *J. sci. Instrum.* **38**, 109.
Bloxham, T. W. (1962). *J. sci. Instrum.* **37**, 387.
Boltwood, B. B. (1907). *Amer. J. Sci.* (4) **23**, 77.
Bowie, S. H. U. (1954). "Nuclear Geology", ed. by H. Faul, p. 48. Wiley, New York.
Brewer, A. K. (1936). *J. chem. Phys.* **4**, 350.
Brewer, A. K. (1938a). *Industr. Engng Chem.* **30**, 893.
Brewer, A. K. (1938b). *J. Amer. chem. Soc.* **60**, 691.
Broecker, W. S., Olson, E. A. and Bird, J. (1959). *Nature* **183**, 1582.
Bullard, E. C. and Stanley, J. P. (1949). Suomen Geodeettisen Laitoksen Julkaisuja; Veröffentl. Finnisch Geodät. Inst. No. 36, p. 33.
Burke, W. H. and Meinschein, W. G. (1955). *Rev. sci. Instrum.* **26**, 1137.
Burst, J. F. (1958). *Amer. Min.* **43**, 481.
Cahen, L., Eberhardt, P., Geiss, J., Houtermans, F. G., Jedwab, J. and Signer, P. (1958). *Geochim. cosmochim. Acta* **14**, 134.
Campbell, N. R. and Wood, A. (1906). *Proc. Camb. phil. Soc.* **14**, S, 15.
Cannon, R. S., Pierce, A. P. and Delevraux, M. H. (1963). *Science* **142**, No. 3592, 574.
Cannon, R. S., Stieff, L. R. and Stern, T. W. (1958). Proc. U.N. 2nd Int. Conf. on Peaceful Uses Atomic Energy. Geneva, Vol. 2, P/773, p. 215.

Cannon, R. S., Pierce, A. P., Antweiler, J. C. and Buck, K. L. (1961). *Econ. Geol.* **56**, No. 1, 1.

Cannon, R. S., Pierce, A. P., Antweiler, J. C. and Buck, K. L. (1962). *Bull. geol. Soc. Amer.*, A. F. Buddington Volume, 115.

Carr, D. R. and Kulp, J. L. (1955). *Rev. sci. Instrum.* **26**, 379.

Carr, D. R. and Kulp, J. L. (1957). *Bull. geol. Soc. Amer.* **68**, 763.

Carslaw, N. S. and Jaeger, J. C. (1947). "Conductions of Heat in Solids". Oxford University Press, London.

Catanzaro, E. J. and Gast, P. W. (1962). *Geochim. cosmochim. Acta* **19**, 113.

Catanzaro, E. J. and Kulp, J. L. (1964). *Geochim. cosmochim. Acta* **28**, 87.

Chaundhury, F. and Sen, K. (1942). *Proc. nat. Inst. Sci. India* **8**, 45.

Cheng, K. L. (1958). *Analyt. Chem.* **30**, No. 6, 1027.

Cherdyntsev, V. V., Oslov, D. P., Isabaev, E. N. and Ivanov, V. I. (1961). *Geokhimiya* 840.

Chow, T. J. and McKinney, C. R. (1958). *Analyt. Chem.* **30**, 1499.

Chow, T. J. and Patterson, C. (1959). *Geochim. cosmochim. Acta* **17**, 21.

Chow, T. J. and Patterson, C. (1962). *Geochim. cosmochim. Acta* **26**, 263.

Clarke, W. B. and Thode, H. G. (1963). *In* "Isotopic and Cosmic Chemistry", ed. by H. Craig, S. L. Miller and G. L. Wasserburg, p. 471. North Holland Publ. Co., Amsterdam.

Clusius, K., Schumacher, E., Hurzeler, N. and Hosteltler, A. U. Z. (1956). *Z. Naturf.* **11a**, 709.

Cobb, J. C. (1961). *Ann. N.Y. Acad. Sci.* **91**, Art. 2, 311.

Cobb, J. C. (1964). *J. geophys. Res.* **69**, No. 9. 1895.

Cobb, J. C. and Kulp, J. L. (1961). *Geochim. cosmochim. Acta* **24**, 226.

Cohen, K. (1951). "The Theory of Isotope Separation". McGraw-Hill, New York.

Collins, C. B., Farquhar, R. M. and Russell, R. D. (1954). *Bull. geol. Soc. Amer.* **65**, 1.

Collins, C. B., Russell, R. D. and Farquhar, R. M. (1953). *Canad. J. Phys.* **31**, 402.

Compston, W. and Jeffery, P. M. (1960). *Nature* **184**, 1792.

Compston, W. and Jeffery, P. M. (1961). *Ann. N.Y. Acad. Sci.* **91**, 185.

Compston, W. and Pidgeon, R. T. (1962). *J. geophys. Res.* **67**, No. 9, 3493.

Compston, W., Jeffery, P. M. and Riley, G. H. (1960). *Nature* **186**, 702.

Cooper, J. A. (1963). *Geochim. cosmochim. Acta* **27**, 525.

Craig, H. (1953). *Geochim. cosmochim. Acta* **3**, 53.

Craig, H. (1954). *J. Geol.* **62**, 115.

Craig, H. (1957). *Tellus* No. 1, **9**, 1.

Craig, H. (1961). *Radiocarbon* **3**, 1.

Craig, R. D. (1956). *J. sci. Instrum.* **36**, 38.

Crane, H. R. (1951). *Nucleonics* No. 6, **9**, 16.

Crane, H. R. (1954). Andover Conf. on Radiocarbon Dating, Phillips Academy, October 21–23.

Crouch, E. A. C. and Webster, R. K. (1963). *J. chem. Soc.* **18**, 118.

Crozaz, G., De Breuk, W. and Picciotto, E. (1964). *J. geophys. Res.* **69**, 2597.

Curran, S. C., Dixon, D. and Wilson, H. W. (1951). *Phys. Rev.* **84**, 151.

Curran, S. C., Dixon, D. and Wilson, H. W. (1952). *Phil. Mag.* **43**, 82.

Curtis, G. H. and Evernden, J. F. (1962). *Nature* **194**, 611.

Curtis, G. H. and Reynolds, J. H. (1958). *Bull. geol. Soc. Amer.* **69**, 151.

Curtis, G. H., Lipson, J. and Evernden, J. F. (1956). *Nature* **178**, No. 4546, 1360.

Cuvier, G. and Brongniart, P. (1808). *J. Mines* XXIII, 421.

Damon, P. E. (1954). *Trans. Amer. geophys. Un.* **35**, 631.

Damon, P. E. and Bikerman, M. (1964). *Bull. geol. Soc. Amer.* **76**, 269.

Damon, P. E. and Green, W. D. (1963). *In* "Radioactive Dating", p. 55. IAEA, Vienna.
Damon, P. E. and Kulp, J. L. (1957). *Amer. Min.* **43**, 433; *Trans. Amer. geophys. Un.* **38**, No. 6, 945.
Damon, P. E. and Kulp, J. L. (1958). *Amer. Min.* **43**, 433.
Dana, J. D. (1880). "Manual of Geology".
Davidson, C. F. (1953). *Min. Mag.* **88**, 73.
Davis, G. L. (see Carnegie Inst. Year Book, 1954).
Davis, G. L. and Aldrich, L. T. (1953). *Bull. geol. Soc. Amer.* **64**, 379.
Davis, R. and Schaeffer, O. A. (1955). U.S. Atomic Energy Comm. Rept. B.N.L. 340.
Dean, J. A. (1960). "Flame Photometry". McGraw-Hill, New York.
Decat, D., van Zanten, B. and Leliaert, G. (1963). *Analyt. Chem.* **35**, 845.
Deffeyes, K. S. and Martin, E. L. (1962). *Science* **136**, 782.
Deuser, W. G. and Herzog, L. F. (1963). *Trans. Amer. geophys. Un.* **44**, No. 1, 111.
Deutsch, S., Hirschberg, D. and Picciotto, E. (1956). *Bull. Soc. belge Géol. Pal. Hydr.* **65**, 267.
Doe, B. R. (1962). *Bull. geol. Soc. Amer.* **73**, 833.
Doe, B. R. and Hart, S. R. (1963). *J. geophys. Res.* **68**, No. 11, 3521.
Dorn, T. F., Fairhall, A. W., Schell, W. R. and Takashima, Y. (1962). *Radiocarbon* **4**, 1.
Duckworth, H. E. (1958). "Mass Spectroscopy." Cambridge University Press.
Dyer, F. F., Emery, J. F. and Leddicotte, G. W. (1962). O.R.N.L. Rept. 3342. UC-4-Chemistry TID-4500 (17th Ed., Rev.).
Easton, A. J. and Lovering, J. F. (1963). *Anal. chim. Acta* **30**, 543.
Eberhardt, P. and Geiss, J. (1960a). *Z. Naturf.* **15a**, 547.
Eberhardt, P. and Geiss, J. (1960b). Varenna Summer School. Nuclear Geology.
Eberhardt, P. and Geiss, J. (1963). *In* "Isotopic and Cosmic Chemistry", ed. by H. Craig, S. L. Miller and G. J. Wasserburg, p. 452. North Holland Publ. Co., Amsterdam.
Eberhardt, P., Delwiche, R. and Geiss, J. (1964). *Z. Naturf.* **19**, 6, 736.
Eberhardt, P., Houtermans, F. G. and Signer, P. (1962), *Geol. Rdsch.* **52**, 836.
Eberhardt, P., Geiss, J., Houtermans, F. G., Buser, W. and von Gunton, H. R. (1955). Atti del 1° Convegno di Geologia Nucleare Roma, p. 149.
Eckelmann, W. R. and Kulp, J. L. (1956). *Bull. geol. Soc. Amer.* **67**, 35.
Eckelmann, W. R. and Kulp, J. L. (1957). *Bull. geol. Soc. Amer.* **68**, 1117.
Eckhoff, H. J. (1960). *Appl. Spectrosc.* **14**, No. 3, 74.
Ecko, M. W. and Turk, E. H. (1957). PTR-143, (Jan. 28).
Edwards, G. and Urey, H. C. (1955). *Geochim. cosmochim. Acta* **7**, No. 314, 154.
Egelkraut, K. and Leutz, H. (1961). *Z. Phys.* **161**, 13.
Ehman, W. D. (1957). U.S. Atomic Energy Comm. NYO-6634.
Ehrenberg, H. Fr., Geiss, J. and Taubert, R. (1955). *Z. angew. Phys.* **7**, 416.
Eicholz, G. G., Hilborn, J. W. and McMahon, C. (1953). *Canad. J. Phys.* **31**, 613.
Eklund, S. (1946). *Ark. Mat. Astr. Fys.* **33a**, No. 14.
Elasser, W., Ney, E. P. and Winckler, J. R. (1956). *Nature* **178**, 1226.
Emiliani, C. (1955). *J. Geol.* **63**, 538.
Erickson, G. P. and Kulp, J. L. (1961). *Bull. geol. Soc. Amer.* **72**, 649.
Evans, C. (1963). D.Phil. Thesis, Oxford.
Evernden, J. F. and Richards, J. R. (1961). *J. geol. Soc. Aust.* **9**, 1.
Evernden, J. F., Curtis, G. H. and Kistler, R. W. (1957). *Quaternaria* **4**, 13.
Evernden, J. F., Curtis, G. H., Kistler, R. W. and Obradovich, J. (1960). *Amer. J. Sci.* **258**, 583.
Ewald, H. and Hintenberger, H. (1953). "Methoden und Anwendungen der Massenspektroskopie." Verlag Chemie, Weinheim.
Facchini, U., Forte, M., Malvicini, A. and Rossini, T. (1956). *Nucleonics* **14**, 126.

Fairbairn, H. W., Faure, G., Pinson, W. H., Hurley, P. M. and Powell, J. L. (1963). *J. geophys. Res.* **68**, No. 24, 6515.

Fanale, F. P. and Kulp, J. L. (1962). *Econ. Geol.* **57**, 735.

Farquhar, R. M. and Cumming, G. L. (1954). *Trans. R. Soc. Canad.* **48**, Ser. 3, 9.

Farquhar, R. M. and Russell, R. D. (1963). *Geochim. cosmochim. Acta* **27**, 1143.

Farrar, E., Macintyre, R. M., York, D. and Kenyon, W. J. (1964). *Nature* **204**, 531.

Faul, H. (1960). *Bull. geol. Soc. Amer.* **71**, 637.

Faure, G. (1963). M.I.T. Ann. Progr. Rep. Contract AT(30–1)–1381, No. 11, p. 125.

Faure, G. and Hurley, P. M. (1963). *J. Petr.* **4**, 31.

Fechtig, H., Gentner, W. and Kalbitzer, S. (1961). *Geochim. cosmochim. Acta* **25**, No. 4, 297.

Feely, H. W. (1960). *Science* **131**, 645.

Feit, W. (1930). *Z. angew. Chem.* **43**, 459.

Feit, W. (1933). *Z. angew. Chem.* **46**, 216.

Fergusson, G. J. (1953). *N.Z. J. Sci. Tech.* B**35**, 90.

Fergusson, G. J. (1955). *Nucleonics* **13**, No. 1, 18.

Fernald, A. T. (1962). U.S. Geol. Surv. Prof. P. Art. II. 450-B, 29.

Ferrara, G., Ledent, D. and Stauffer, H. (1958). *Com. naz. Ric. nucl. Roma* **1**, 1.

Fireman, E. L. (1958). *Nature* **181**, 1613, 1725.

Fish, R. A. and Goles, G. G. (1962). *Nature* **196**, 27.

Fleischer, M. (1955). Econ. Geol. 50th Anniv. Vol. 970.

Fleischer, R. L. and Price, P. B. (1963a). *J. Geophys. Res.* **68**, No. 16, 4847.

Fleischer, R. L. and Price, P. B. (1963b). General Electric Res. Lab. Rept. 63-RL-3501M.

Fleischer, R. L. and Price, P. B. (1964). *Geochim. cosmochim. Acta* **28**, 1705.

Fleischer, R. L., Price, P. B. and Walker, R. M. (1964), *J. geophys. Res.* **69**, No. 22, 4885.

Fleischer, R. L., Price, P. B., Symes, E. M. and Miller, D. S. (1963). General Electric Res. Lab. Rep. 63RL 3437M.

Fleischer, R. L., Naeser, C. W., Price, P. B., Walker, R. M. and Marvin, U. B. (1965). General Electric Res. Lab. Rep. 65-RL-3901M.

Fleming, E. H., Ghioroso, A. and Cunningham, B. B. (1952). *Phys. Rev.* **88**, 642.

Flinta, J. and Eklund, E. (1952). *Ark. Fys.* **7**, 401.

Flynn, K. F. and Glendenin, L. E. (1959). *Phys. Rev.* **116**, 744.

Folinsbee, R. E., Lipson, J. and Reynolds, J. H. (1956). *Geochim. cosmochim. Acta* **10**, 60.

Folinsbee, R. E., Baadsgaard, H. and Lipson, J. (1961). *Ann. N.Y. Acad. Sci.* **91**, Art. 2, 352.

Fornaseri, F. and Grandi, L. (1960). *Geochim. cosmochim. Acta* **19**, 218.

Fritze, K. and MacMullin, C. C. (1964). Personal communication.

Fritze, K. and Strassmann, F. (1956). *Z. Naturf.* **11a**, 277.

Galliher, E. W. (1935). *Bull. geol. Soc. Amer.* **46**, 1351.

Garrels, R. M. (1941). *Econ. Geol.* **36**, 848.

Gast, P. W. (1960a). *J. geophys. Res.* **65**, 1287.

Gast, P. W. (1960b). *Geochim. cosmochim. Acta* **19**, 1.

Gast, P. W. (1961). *Ann. N.Y. Acad. Sci.* **91**, Art. 2, 181.

Gast, P. W. (1962). *Geochim. cosmochim. Acta* **26**, 927.

Gastil, G. (1960). *Amer. J. Sci.* **258**, 1.

de Geer, G. (1912). C.R. XI. Congr. géol. int. Stockholm 1910, p. 241.

Geese-Bähnisch, I. and Huster, E. (1954). *Naturwissenschaften* **41**, 495.

Geese-Bähnisch, I., Huster, E. and Walcher, N. (1952). *Naturwissenschaften* **39**, 379.

Gefeller, C. and Oeschger, H. (1962). *Helv. phys. Acta* **XXXV**. 307.

Gehrke, E. and Reichenheim, O. (1906). *Verh. dtsch. phys. Ges.* **8**, 559.

Gehrke, E. and Reichenheim, O. (1907). *Verh. dtsch. phys. Ges.* **9**, 76, 200, 376.

Geikie, A. (1822). Presidential Address, British Association, Edinburgh.

Geiss, J. (1954). *Z. Naturf.* **99**, 218.

Geiss, J., Gefeller, C., Houtermans, F. G. and Oeschger, H. (1958). U.N. Peaceful Uses Atomic Energy, 2nd Ed., P/236, 21, p. 147.

Gentner, W. and Trendellenberg, E. (1954). *Geochim. cosmochim. Acta* **6**, 26.

Gerling, E. K. (1942). *Dokl. Akad. Nauk SSSR* **34**, 259.

Gerling, E. K. (1960). "Sovremenneye sostoyaniye argonovogo metoda opredeleniya vozrasta." Akademia Nauk. USSR, Moscow.

Gerling, E. K. and Pavlova, T. G. (1951). *Dokl. Akad. Nauk SSSR* **77**, 85.

Gerling, E. K. and Polkanov, A. A. (1958). *Geokhimiya* **8**, 4.

Gerling, E. K., Morozova, I. M. and Kurbatov, V. V. (1961a) *Ann. N.Y. Acad. Sci.* **91**, Art. 2, 227.

Gerling, E. K., Morozova, I. M. and Kurbatov, V. V. (1961b). *Geokhimiya* No. 1, 45.

Gerling, E. K., Yermolin, G. M., Baranovskaya, N. V. and Titov, N. E. (1952). *Dokl. Akad. Nauk SSSR* **86**, 593.

Gibson, W. M. (1961). Nat. Acad. Sci. Nat. Res. Coun. Nucl. Sci. Ser. NAS–NS–3040.

Giletti, B. J. and Kulp, J. L. (1955). *Amer. Min.* **40**, 481.

Giletti, B. J., Bazan, F. and Kulp, J. L. (1958). *Trans. Amer. geophys. Un.* **39**, No. 5, 166.

Glendenin, L. E. (1961). *Ann. N. Y. Acad. Sci.* 91, Art. 2, 166.

Glock, W. S. and Agerter, S. (1963). *Endeavour* **22**, 9.

Goldberg, E. D. (1963). Symp. Radioactive Dating Athens. Nov. 1962, p. 121. IAEA, Vienna.

Goldberg, E. D. and Koide, M. (1962). *Geochim. cosmochim. Acta* **26**, 417.

Goldberg, E. D., Koide, M., Schmitt, R. A. and Smith, R. H. (1963). *J. geophys. Res.* **68**, No. 14.

Goldschmidt, V. M. (1929). *Naturwissenschaften* **17**, 134.

Goldschmidt, V. M. (1937). *Skr. norsk. Videnk. Akad. Math. Nat. Kl.* No. 4, 140.

Goldschmidt, V. M. (1954). "Geochemistry". Oxford University Press, London.

Goodchild, J. G. (1897). *Proc. R. phys. Soc. Edinb.* **14**, 259.

Goodman, C. and Evans, R. D. (1941). *Bull. geol. Soc. Amer.* **52**, 491.

Goodwin, H. (1962). Dating Conference, Cambridge University, England; *Nature* **195**, 943, 984.

Goris, P. (1962). Personal communication.

Gottfried, D., Jaffe, H. W. and Senftle, F. E. (1959). *Bull. U.S. geol. Survey* **1097–A**, 1.

Gottfried, D., Senftle, F. E. and Waring, C. L. (1956). *Amer. Min.* **41**, 157.

Greenhalgh, D. and Jeffery, P. M. (1959). *Geochim. cosmochim. Acta* **16**, 39.

Grimaldi, F. S. (1952). *U.S. geol. Surv. Circ.* **199**, 20.

Grimaldi, F. S. and Fletcher, M. H. (1956). *Analyt. Chem.* **28**, 812.

Grimaldi, F. S. and Jenkins, L. F. (1957). *Analyt. Chem.* **29**, 848.

Grimaldi, F. S., May, I., Fletcher, M. H. and Titcomb, I. (1954). *Bull. U.S. geol. Surv.* **1006.**

Grindler, J. E. (1962). "The Radiochemistry of Uranium". Nat. Acad. Sci. Nat. Res. Council. Nat. Sci. Ser. NAS-NS 3050.

Groves, A. W. (1951). "Silicate Analysis", p. 77. Murby, London.

Grummitt, W. E., Brown, R. M., Cruickshank, A. J. and Fowler, I. L. (1956). *Canad. J. Chem.* **34**, 206.

Haber, F. C. (1959). "The Age of the World, Moses to Darwin". Johns Hopkins Press, Baltimore.

Hagemann, F. (1950). *J. Amer. chem. Soc.* **72**, 768.

Hagemann, F., Gray, J. G., Machta, L. and Turkevich, A. (1959). *Science* **130**, 542.

Hahn, O. and Rothernback, M. (1919). *Z. Phys.* **20**, 194.

Hahn, O., Strassmann, F. and Walling, E. (1937). *Naturwissenschaften* **25**, 189.

Hales, A. L. (1961). *Ann. N.Y. Acad. Sci.* **91**, Art. 2, 524.

Hamaguchi, H., Reed, G. W. and Turkevich, A. (1957). *Geochim. cosmochim. Acta* **12**, 337.

Hamilton, E. I. (1959a). *Geol. Mag.* XCVII, No. 3, 255.

Hamilton, E. I. (1959b). *Medd. om. Grønland* **162**, No. 7, 1.

Hamilton, E. I. (1960). *Medd. om. Grønland* **162**, No. 8, 1.

Hamilton, E. I. (1963). *J. Petr.* **4**, Pt. 3, 383.

Hamilton, E. I. (1964). *Medd. om. Grønland* **162**, No. 10, 1.

Hamilton, E. I. (1965). *Nature* **206**, 251.

Hamilton, E. I. and Deans, T. (1963). *Nature* **198**, No. 4882, 776.

Hamilton, E. I., Dodson, M. H. and Snelling, N. J. (1962). *Int. J. appl. Radn Isotopes* **13**, 587.

Haring, A., de Vries, A. E. and de Vries, H. (1958). *Science* **128**, 472.

Hart, S. R. (1961). *J. geophys. Res.* **66**, No. 9, 2995.

Hart, S. R. and Dodd, R. T. (1962). *J. Geophys. Res.* **67**, No. 7, 2998.

Hart, S. R., Aldrich, L. T., Davis, G. L., Tilton, G. R., Baadsgaard, H., Kouvo, O. and Steiger, R. N. (1963). *Carnegie Inst. Wash. Ann. Rep.* Year Book 62, 267.

Hawley, J. E. and Nichol, I. (1961). *Econ. Geol.* **56**, No. 3, 467.

Haxel, O., Houtermans, F. G. and Kemmerich, M. (1948). *Phys. Rev.* **74**, 1886.

Hayes, F. N. (1955). *I.R.E. Trans. Prof. Gp. Nuc. Sci.* NS-5. No. 3, 166.

Hayes, F. N. (1956). *Int. J. Appl. Rad. Isotopes* **1**, 46.

Hayes, F. N., Anderson, E. C. and Arnold, J. R. (1953). U.N. Proc. Peaceful Uses Atomic Energy, Geneva. XIV, P/68, 188.

Hayes, F. N., Anderson, E. C. and Arnold, J. R. (1956). Proc. Int. Conf. Peaceful Uses Atomic Energy 1. U.N. Pub. IX 1, 14, 188.

Hedge, C. E. and Walthall, F. G. (1963). *Science* **140**, No. 3572, 1214.

Heier, K. S. (1964). *Nature* **202**, 477.

Heier, K. S. and Adams, J. A. S. (1964). *Phys. Chem. Earth* **5**, 253.

Heier, K. S., McDougall, I. and Adams, J. A. S. (1964). *Nature* **201**, No. 4916, 254.

Helgeson, H. C. (1964). Int. Ser. Monographs on Earth Sciences, ed. by D. E. Ingerson. Vol. 17, "Complexing and hydrothermal ore deposition", p. 128.

Hemmendinger, A. and Smythe, W. R. (1937). *Phys. Rev.* **51**, 1052.

Henderson, E. P. and Clarke, R. S. (1962). Program. Abst. Amer. Geophys. Union Meeting, April, 1962, p. 106.

Herr, W. and Merz, E. (1958). *Z. Naturf.* **13a**, 231.

Herr, W., Hoffmeister, W. and Langhoff, J. (1960). *Z. Naturf.* **15a**, 99.

Herrmann, R. and Alkemade, C. Th. (1960). "Flammenphotometrie", 2nd Ed. Springer, Berlin.

Herzog, L. F. and Pinson, W. H. (1956). *Amer. J. Sci.* **254**, 555.

Hester, J. H. (1960–1). *Amer. Antiquity* **26**, 58.

Heyden, M. and Kopfermann, H. (1938). *Z. Phys.* **108**, 232.

Hillebrand, W. F., Lundell, G. E. F., Bright, H. A. and Hoffman, J. H. (1953). "Applied Inorganic Analysis", 2nd Ed. Wiley, New York.

Hintenberger, H. (ed.) (1957). "Nuclear Masses and their Determination". Pergamon Press, London.

Hintenberger, H. (1962). *Ann. Rev. nucl. Sci.* **12**, 477.

Hirt, B., Tilton, G. R., Herr, W. and Hoffmeister, W. (1963). *In* "Earth Science and Meteoritics", ed. by J. Geiss and E. D. Goldberg, p. 273. North Holland Publ. Co., Amsterdam.

Hoff, L. U. (1938). *Phys. Rev.* **853**, 845.

Hoffman, J. H. and Nier, A. O. (1958). *Phys. Rev.* **112**, 2112.

Hoffman, J. H. and Nier, A. O. (1959). *Geochim. cosmochim. Acta* **17**, 32.

Hoffman, K. W. (1961). *Naturwissenschaften* **48**, 36.

Holland, H. D. and Gottfried, D. (1955). *Acta cryst.* **8**, 291.

Holland, H. D. and Kulp, J. L. (1950). *Science* **111**, 312.

Holland, H. D. and Kulp, J. L. (1954). *Geochim. cosmochim. Acta* **5**, 197.

Holmes, A. (1913). "The Age of the Earth". Harper Brothers, New York.

Holmes, A. (1929). *Nature* **124**, 477.

Holmes, A. (1931). *Trans. geol. Soc. Glasgow* (for 1928–1929), **18**, 559.

Holmes, A. (1932). *Geol. Mag.* LXIX, No. 822, 543.

Holmes, A. (1937). *Econ. Geol.* **32**, 763.

Holmes, A. (1946). *Nature* **157**, 680.

Holmes, A. (1947). *Trans. geol. Soc. Glasgow* **21**, 117.

Holmes, A. (1948). *Trans. Edinb. geol. Soc.* **14**, 11.

Holmes, A. (1955). *Proc. geol. Ass. Canada* **7**, 81.

Holmes, A. (1960). *Trans. Edinb. geol. Soc.* **17**, 183.

Holyk, W. K. (1952). Ph.D. Thesis. Massachusetts Institute of Technology.

Horberg, L. (1955). *J. Geol.* **63**, 278.

Horstman, E. L. (1956). *Analyt. Chem.* **28**, 1417.

Houtermans, F. G. (1946). *Naturwissenschaften* **6**, 185.

Houtermans, F. G. (1947). *Z. Naturf.* **2a**, 322.

Houtermans, F. G. (1953). *Nuovo Cimento* **10**, No. 12, 1623.

Houtermans, F. G. (1960). *Geol. Rdsch.* **49/1**, 168.

Houtermans, F. G. and Eberhardt, P. (1960). Summer Course Nuclear Geology, Varenna.

Houtermans, F. G., Eberhardt, P. and Ferrara, G. (1963). "Isotopic and Cosmic Chemistry", p. 233. North Holland Publ. Co., Amsterdam.

Hower, J. (1961). *Amer. Min.* **46**, 313.

Hower, J., Hurley, P. M., Pinson, W. H. and Fairbairn, H. W. (1963). *Geochim. cosmochim. Acta* **27**, 405.

Hurley, P. M. (1956). *Bull. geol. Soc. Amer.* **67**, 395.

Hurley, P. M. (1959). "How Old Is the Earth?" The Science Study Series. Heinemann, London.

Hurley, P. M. (1961). *Ann. N.Y. Acad. Sci.* **91**, Art. 2, 294.

Hurley, P. M. (1963). *Tech. Rev.* November, 28.

Hurley, P. M. and Fairbairn, H. W. (1953). *Bull. geol. Soc. Amer.* **64**, 659.

Hurley, P. M. and Fairbairn, H. W. (1957). *Trans. Amer. geophys. Un.* **38**, 936.

Hurley, P. M., Larsen, E. S. and Gottfried, D. (1956). *Geochim. cosmochim. Acta* **9**, 98.

Hurley, P. M., Fairbairn, H. W., Faure, G. and Pinson, W. H. (1963). "Radioactive Dating". p. 201. IAEA, Athens.

Hurley, P. M., Hunt, J. M., Pinson, W. H. and Fairbairn, H. W. (1963). *Geochim. cosmochim. Acta* **27**, 279.

Hurley, P. M., Hughes, T. C., Faure, G., Fairbairn, H. W. and Pinson, W. H. (1962). *J. geophys. Res.* **67**, 5315.

Hybbinette, A. G. (1943). *Svensk. Kem. Tidskr.* **55**, 151.

Hyde, E. K. (1960). *Univ. Calif. Rep.* UCEL 9036.

Ingram, M. G. (1948). *In* "Advances in Electronics and Electron Physics", Vol. 1, ed. by L. Marton. Academic Press, New York.

Ingram, M. G. (1953). *Nat. Bur. Stand. Circ.* **522**, 204.

Inghram, M. G. and Chupka, W. A. (1953). *Rev. sci. Instrum.* **24**, No. 7, 518.

Inghram, M. G. and Hayden, R. J. A. (1954). "A Handbook on Mass Spectroscopy", Nat. Acad. Sci. Nat. Res. Council. Pub. 311, p. 37.

Isabaev, E. N., Usatov, E. P. and Cherdyntsev, V. V. (1960). *Radiokhimiya* **2**, 94.

Jäger, E. (1962). *J. geophys. Res.* **67**, No. 13, 5293.

Jäger, E., Niggli, E. and Baethe, N. (1963). *Schweiz. Min. Petr. Mitt.* 43/2, 465.

Jamieson, R. T. and Schreiner, G. D. L. (1957). *Proc. roy. Soc.* **B146**, 257.

Jeffrey, P. M. and Reynolds, J. H. (1961). *J. geophys. Res.* **66**, 3582.

Jenkins, E. N. (1955). *Analyst* **80**, 301.

Jenkins, E. N. and Smales, A. A. (1956). *Quart. Rev. chem. Soc.* X, No. 1, 83.

Jenkins, L. (1954). U.S. Atomic Energy Comm. TE1, 453.

Jensen, M. L. (1957). *Econ. Geol.* **52**, 269.

Jensen, M. L. (1959). *Econ. Geol.* **54**, 374.

Joly, J. (1899). *Sci. Trans. R. Dublin Soc.* (2) **7**, 23.

Joly, J. (1907). *Phil. Mag.* **13**, 381.

Joly, J. (1923). *Sci. Monthly* **16**, 205.

Joly, J. and Rutherford, E. (1913). *Phil. Mag.* **25**, 644.

Jones, W. M. (1955). *Phys. Rev.* **100**, 124.

Jost, W. (1952). "Diffusion in Solids, Liquids and Gases".

Kanasewich, E. R. (1962). *Geophys. J. Roy. Astr. Soc.* **1**, 158.

Kanasewich, E. R. and Slawson, W. F. (1964). *Geochim. cosmochim. Acta* **28**, 541.

Karamyan, K. A. (1962). *Geokhimiya* No. 2, 191.

Katz, J. J. and Rabinowitch, E. (1951). "The Chemistry of Uranium".

Kaye, M. M., Nicolaysen, L. O., Willis, J. P. and Ahrens, L. H. (1965). *Geochim. cosmochim. Acta* (in press).

Keer, P. F. and Kulp, J. L. (1952). *Science* **115**, 86.

Keevil, N. B. (1939). *Amer. J. Sci.* **237**, 195.

Kember, N. F. (1952). *Analyst* **77**, 78.

Kemmerich, M. (1949). *Z. Phys.* **126**, 399.

Kendall, B. R. F. (1958). *Rev. sci. Instrum.* **29**, No. 10, 851.

Khlopin, V. G. and Gerling, E. K. (1947). *Dokl. Akad. Nauk SSSR* **58**, 1415.

King, H. F. and Thompson, B. P. (1953). 5th Empire Mining and Metallurgical Congress, Vol. 1, p. 533.

Kinser, C. A. (1954). *U.S. Geol. Surv. Circ.* 330.

Kirsten, T., Krankowsky, D. and Zähringer, J. (1963). *Geochim. cosmochim. Acta* **27**, 13.

Knopf, A. F., Schuchert, C., Kovarik, A. F., Holmes, A. and Brown, E. W. (1931). *Bull. nat. Res. Coun.* **5**, No. 80.

Koczy, F. F. (1949). *Geol. Förening Stockh. Förhändl.* **71**, 238.

Koczy, F. F. (1958). Proc. 2nd U.N. Int. Conf. P.U.A.E. **18**, 351.

Koczy, F. F. (1963). *In* "The Sea", ed. by M. N. Hill, Chap. 30. Wiley, New York.

Koczy, F. F. and Titze, F. (1958). *J. Marine Res.* **17**, 302.

Koenigswald, G. H. R., Gentner, W. and Lippolt, H. J. (1961). *Nature* **192**, No. 4804, 720.

Köhlhorster, J. (1930). *Z. Geophys.* **6**, 341.

Kohman, T. P. (1956). *Ann. N.Y. Acad. Sci.* **62**, 505.

Kohman, T. P. and Goel, P. S. (1962). *In* "Radioactive Dating", pp. 72, 148. IAEA, Vienna.

Kohman, T. P. and Saito, N. (1954). *Ann. Rev. nucl. Sci.* **4**, 401.

Komlev, L. V. (1958). *Geokhimiya* **7**, 121.

König, G. and Wänke, H. (1950). *Z. Naturf.* **14a**, 866.

Korff, S. A. (1940). *Terr. Magn. atmos. Elect.* **45**, 133.

Kraus, K. A. and Nelson, F. (1956). Int. Conf. Peaceful Uses Atomic Energy, P/837. Sess. 98.

Krauskopf, K. (1955). *Geochim. Cosmochim. Acta* **9**, 1.

Kroll, V. (1954). *Deep-sea Res.* **1**, 211.

Kroll, V. (1955). *Rep. Swed. Deep Sea Expedition* **10**, 1. Göteborg.

Krummenacher, D., Merrihue, C. M., Pepin, R. O. and Reynolds, J. H. (1962). *Geochim. cosmochim. Acta* **26**, 231.

Krylov, A. Ya. (1961). *Ann. N.Y. Acad. Sci.* **91**, Art. 2, 324.

Krylov, A. Ya. and Silin, Y. I. (1959). *Dokl. Akad. Nauk SSSR* **129**, 3.

Krylov, A. Ya. and Silin, Y. I. (1960). *Izv. Akad. Nauk. SSSR* Ser. Geol. 1, 56.

Krylov, A. Ya., Baranovskaya, N. V. and Lovtsyus, G. P. (1958). Trans. 5th Session Comm. Abs. Dating Geol. Form, p. 254.

Krylov, A. Ya., Baranovskaya, N. V., Lovtsyus, G. P., Drozhzhin, V. M. and Litvina, L. A. (1958). Trans. 5th Sess. CADGF, p. 254.

Kulp, J. L. (1952). *Sci. Monthly* **75**, 259.

Kulp, J. L. (1953). *Atomics* 4.

Kulp, J. L. (1955). Geol. Soc. Amer. Spec. Paper 62, p. 609.

Kulp, J. L. (1960). Int. Geol. Cong. 21st Copenhagen, Proc. Pt. 3, p. 18.

Kulp, J. L. (1961). *Science* **133**, No. 3459, 1105.

Kulp, J. L. and Eckelmann, W. R. (1957). *Amer. Min.* **42**, 154.

Kulp, J. L. and Engels, J. (1963). *In* "Radioactive Dating", p. 219. IAEA, Vienna.

Kulp, J. L. and Volchok, H. L. (1953). *Phys. Rev.*

Kulp, J. L., Ault, W. U. and Feely, H. W. (1956). *Econ. Geol.* **51**, 139.

Kulp, J. L., Bate, G. L. and Broecker, W. S. (1954). *Amer. J. Sci.* **252**, 345.

Kulp, J. L., Broecker, W. S. and Eckelmanr, W. R. (1953). *Nucleonics* **11**, No. 8, 19.

Kulp, J. L., Feely, H. W. and Tryon, L. E. (1951). *Science* **114**, 565.

Kulp, J. L., Volchok, H. L. and Holland, H. D. (1952a). *Trans. Amer. geophys. Un.* **33**, 101.

Kulp, J. L., Volchok, H. L. and Holland, H. D. (1952b). *Amer. Min.* **37**, 709.

Kuroda, P. K. (1955). *Ann. N.Y. Acad. Sci.* **62**, 117.

Kuroda, P. K. (1961). *Geochim. Cosmochim. Acta* **24**, 40.

Kuroda, P. K. (1963). *In* "Radioactive Dating", p. 45. IAEA, Vienna.

Ladenburg, R. (1952). *Phys. Rev.* **86**, 128.

Lal, D., Goldberg, E. D. and Koide, M. (1960). *Science* **131**, No. 3397, 332.

Lane, A. C. (1929). *Amer. J. Sci.* (5), **17**, 342.

Lane, A. C. (1941). Science in Progress. Ser. 2. Yale Univ. Press, p. 107.

Larsen, E. S. (1945). *Amer. J. Sci.* **243**, Daly Vol., 369.

Larsen, E. S. and Cross, W. (1956). U.S. Geol. Survey Prof. Paper 258, 303.

Larsen, E. S. and Keevil, N. B. (1947). *Bull. geol. Soc. Amer.* **58**, 483.

Larsen, E. S., Keevil, N. B. and Harrison, N. C. (1952). *Bull. geol. Soc. Amer.* **63**, 1045.

Leakey, L. S. B., Evernden, J. F. and Curtis, G. H. (1961). *Nature* **191**, No. 4787, 478.

Leddicotte, G. W. (1956). *Nucleonics* **14**, 5, 47.

Leddicotte, G. W. (1961a). Radioactivation Analysis. O.R.N.L. 60–11–124.

Leddicotte, G. W. (1961b). Nat. Acad. of Sci. Nat. Res. Council Nac. Sci. Ser. NAR–NS–3028.

Leddicotte, G. W. (1962). *Analyt. Chem.* **34**, 143R.

Leech, G. B. and Wanless, R. D. (1962). A. Buddlington Vol. *Geol. Soc. Amer.* 241.

Lehmann, J. G. (1767). Versuch einer Geschichte von Flötz-Gebürgen, Berlin.

Leipziger, F. D. and Croft, N. J. (1964). *Geochim. cosmochim. Acta* **28**, 268.
Lessing, P. and Catanzaro, E. J. (1964). *J. geophys. Res.* **69**, No. 4, 1599.
Lessing, P., Decker, R. W. and Reynolds, R. C. (1963). *J. geophys. Res.* **68**, No. 20, 5851.
Leutz, H., Wenninger, H. and Ziegler, K. (1962). *Z. Phys.* **169**, 409.
Levine, H. and Grimaldi, F. S. (1950). *AEC Rep.* AECD-2824.
Lewis, G. M. (1952). *Phil. Mag.* **43**, 1070.
Libby, W. F. (1946). *Phys. Rev.* **69**, 671.
Libby, W. F. (1953). *Proc. nat. Acad. Sci., Wash.* **39**, 245.
Libby, W. F. (1955). "Radiocarbon Dating", 2nd Ed. University of Chicago Press.
Libby, W. F. (1957). *Analyt. Chem.* **2a**, 1566.
Libby, W. F. (1963). *Science* **140**, 278.
Lindner, M. (1953). *Phys. Rev.* **91**, 642.
Lingenfelter, R. E. (1963). *Rev. Geophys.* **1**, No. 1, 35.
Lippolt, H. J. and Gentner, W. (1963). *In* "Radioactive Dating", p. 239. IAEA, Vienna.
Lipson, J. (1956). *Geochim. cosmochim. Acta* **10**, 149.
Lipson, J. (1958). *Bull. geol. Soc. Amer.* **69**, 137.
Livingstone, D. A. (1963). *Geochim. cosmochim. Acta* **27**, No. 10, 1055.
Long, A., Silverman, A. J. and Kulp, J. L. (1960). *Econ. Geol.* **55**, 645.
Long, L. E., Kulp, J. L. and Eckelmann, F. D. (1959). *Amer. J. Sci.* **257**, No. 8. 585.
Longmuir, J. and Kingdon, W. (1925). *Proc. roy. Soc.* **A107**, 61.
Loughlin, G. F. and Koschmann, A. H. (1942). U.S.G.S. Prof. Paper. 200.
Lounsbury, M. (1956). *Canad. J. Chem.* **34**, 259.
Louw, J. D. (1954). *Trans. geol. Soc.* S.A. LVII, 211.
Louw, J. D. (1955). *Nature* **175**, 349.
Lovering, J. F. (1955). *Econ Geol.* 50th Anniv. No. 249.
Lyell, C. (1841). "Elements of Geology". John Murray, London.
Lyon, W. S. (1964). "Guide to Activation Analysis." Van Nostrand, Princeton.
McDougall, I. (1961). *Nature* **190**, No. 4782, 1184.
McDougall, I. (1963). *J. Geophys. Res.* **68**, No. 5, 1535.
McDougall, I. (1964). *Bull. geol. Soc. Amer.* **75**, 107.
McDougall, I. and Green, D. H. (1964). *Norsk Geol. Tidskr.* **44**, Pt. 2, 183.
McGregor, M. H. and Wiedenbeck, M. L. (1952). *Phys. Rev.* **86**, 420.
McGregor, M. H. and Wiedenbeck, M. L. (1954). *Phys. Rev.* **94**, 1938.
Mackenzie, C. (1953). *Min. Mag.* **30**, 22.
McNair, A. and Wilson, H. W. (1961). *Phil. Mag.* **6**, 8th Ser. 563.
Macnamara, J. and Thode, H. G. (1950), *Phys. Rev.* **78**, 307.
Macnamara, J., Fleming, W., Szabo, A. and Thode, H. G. (1952). *Canad. J. Chem.* **30**, 73.
Maddock, A. G. and Willis, E. H. (1961). *Advanc. inorg. Radiochem.* **3**, 287.
Mair, J. A. (1958). Ph.D. Thesis. University of Toronto.
Mair, J. A., Maynes, A. D., Patchett, J. E. and Russell, R. D. (1960). *J. geophys. Res.* **65**, 314.
Mapper, D. (1960). *In* "Methods in Geochemistry", ed. by A. A. Smales and L. R. Wager, Chap. 9. Interscience, New York.
Margoshes, M. and Vallee, B. L. (1956). *Analyt. Chem.* **28**, 180.
Marshall, R. R. (1957). *Geochim. cosmochim. Acta* **12**, 225.
Marshall, R. R. and Feitknecht, J. (1964). *Geochim. cosmochim. Acta* **28**, 365.
Marshall, R. R. and Hess, D. C. (1960). *Analyt. Chem.* **32**, 960.
Marshall, R. R. and Hess, D. C. (1961). *Geochim. cosmochim. Acta* **21**, 161.
Marti, F. B. and Muñoz, J. R. (1957). "Flame Photometry". Elsevier, Amsterdam.
Martin, L. H. and Hill, R. D. (1947). "A Manual of Vacuum Practice". Melbourne University Press.

9

Masuda, A. (1964). *Geochim. cosmochim. Acta* **28**, 291.
Mattauch, J. (1937). *Naturwissenschaften* **25**, 189.
Mattauch, J. (1938). *Z. anorg. Chem.* **236**, 209.
Mattauch, J. (1947). *Angew. Chem.* **A**, No. 2, 37.
Mattauch, J. and Herzog, R. (1934). *Z. Phys.* **89**, 786.
Matthew, W. D. (1914). *Science* N.S. **40**, 232.
Maurette, M., Pellas, P. and Walker, R. M. (1964). *Bull. Soc. franç. Minér. Crist.* LXXXVII, 6.
Miyake, Y. and Sugimura, Y. (1961). *Science* **133**, 1823.
Molyk, A., Drever, W. P. and Curran, S. C. (1955). *Nucleonics* **13**, No. 2, 44.
Moorbath, S. and Bell, J. D. (1965). *J. Petr.* (in press).
Moorbath, S. and Vokes, F. M. (1963). *Norsk. Geol. Tidskr.* **43**, Pt. 3, 283.
Moracherskii, D. E. and Nachaera, A. A. (1960). *Geokhimiya* No. 6, 1960.
Morgan, J. W. and Lovering, J. F. (1963). *Anal. Chim. Acta* **28**, 405.
Morgan, J. W. and Lovering, J. F. (1964). *Science* **144**, No. 3620, 835.
Morrison, P. (1951). *Phys. Rev.* **82**, 209.
Moscicki, W. (1953). *Acta Phys. Polon.* **12**, 238.
Mousuf, A. K. (1952). *Phys. Rev.* **88**, 150.
Mühlhoff, W. (1930). *Ann. Phys.* **1**, 205.
Murray, E. G. and Adams, J. A. S. (1958). *Geochim. cosmochim. Acta* **13**, 260.
Murthy, V. R. (1961). *Phys. Rev. Letters* **5**, 539.
Murthy, V. R. and Patterson, C. C. (1962). *J. geophys. Res.* **67**, 1161.
Murthy, V. R. and Steuber, A. M. (1963). *Trans. Amer. geophys. Un.* 44, No. 1, 112.
Naldrett, S. N. and Libby, W. F. (1948). *Phys. Rev.* **73**, 487.
Naughton, J. J. (1963). *Nature* **197**, No. 4868, 661.
Nesterova, Yu. S. (1958). *Geokhimia* **7**, 835.
Nicolaysen, L. O. (1957). *Geochim. cosmochim. Acta* **11**, 41.
Nicolaysen, L. O. (1961). *Ann. N.Y. Acad. Sci.* **91**, Art. 2, 198.
Nier, A. O. (1935). *Phys. Rev.* **49**, 272.
Nier, A. O. (1938). *J. Amer. chem. Soc.* **60**, 1571.
Nier, A. O. (1939). *Phys. Rev.* **55**, 150.
Nier, A. O. (1940). *Rev. sci. Instrum.* **11**, 212.
Nier, A. O. (1941). *J. appl. Phys.* **12**, 342.
Nier, A. O. (1947). *Rev. sci. Instrum.* **18**, 398.
Nier, A. O. (1950a). *Phys. Rev.* **77**, 789.
Nier, A. O. (1950b). *Phys. Rev.* **79**, 450.
Nier, A. O., Thompson, R. W. and Murphy, B. F. (1941). *Phys. Rev.* **60**, 112.
Noddack, I. and Noddack, W. (1931). *Z. phys. Chem.* **154A**, 207.
Noggle, T. S. and Steigler, J. O. (1960). *J. appl. Phys.* **31**, No. 12, 2199.
Noll, W. (1934). *Chem. d. Erde* **8**, 507.
Oakley, K. P. (1950). *Bull. Brit. Mus. natur. Hist. Geol. Ser.* **2**, 6.
Olsson, I. U. (1957). *Ark. Fys.* **13**, 37.
Olsson, I. U., Karlen, I., Turnbull, A. H. and Prosser, N. J. D. (1962). *Ark. Fys.* **22**, 237.
Orbain, G. (1931). *SB. Akad. Wiss. Wien*, Math. Naturwiss Kl. Abst. **11a**, 140, 121.
Osborn, G. H. and Johns, H. (1951). *Analyst* **76**, 410.
Ostic, R. G. (1963). D. Phil. Brit. Columbia.
Ostic, R. G., Russell, R. D. and Reynolds, P. H. (1963). *Nature* **199**, 1150.
Ovchinnikova, G. V. (1960). *Geokhimiya* **5**, 392.
Pabst, A. (1952). *Amer. Min.* **37**, 137.
Palmer, G. H. (1956). "Electromagnetically Enriched Isotopes and Mass Spectrometry". Butterworths, London.

Palmer, G. H. and Aitkin, K. L. (1953). *J. sci. Instrum.* **30**, 314.

Pandow, M., Mackay, C. and Wolfgang, R. (1960). *J. inorg. nucl. Chem.* **14**, 153.

Paneth, F. and Peters, K. (1928). *Z. phys. Chem.* **134**, 353.

Patterson, C. (1955). *Geochim. cosmochim. Acta* **7**, 141.

Patterson, C. (1956). *Geochim. cosmochim. Acta* **10**, 230.

Patterson, C. (1963). *In* "Isotopic and Cosmic Chemistry", ed. by H. Craig, S. L. Miller and G. J. Wasserburg. North Holland Publ. Co., Amsterdam.

Patterson, C. and Duffield, B. (1963). *Geochim. cosmochim. Acta* **27**, 1180.

Patterson, C. C., Goldberg, E. D. and Inghram, M. G. (1953). *Bull. geol. Soc. Amer.* **64**, Pt. 1, 1387.

Patterson, C. C., Tilton, G. R. and Inghram, M. G. (1955). *Science* **121**, 69.

Patterson, C. C., Brown, H. S., Tilton, G. R. and Inghram, M. (1953). *Phys. Rev.* **92**, 1234.

Peters, B. (1955). *Proc. Indian Acad. Sci.* **41**, 67.

Peters, B. (1957). *Z. Phys.* **148**, 93.

Pettersson, H. (1937). *SB. Akad. Wiss. Wien*, Math. Naturw. K1127. Mitt. Inst. Radiumforsch Wien Nr. 400a.

Picciotto, E. (1950). *Bull. Soc. belge Géol. Pal. Hydrol.* **59**, 170.

Picciotto, E. E. and Deutsch, S. (1960). Summer Course. Nuclear Geology, Varenna.

Picciotto, E. and Wilgain, S. (1954). *Nature* **173**, 632.

Picciotto, E. and Wilgain, S. (1956). *Nuovo Cimento* **4**, 1525.

Picciotto, E., Crozaz, G. and de Breuck, W. (1964). *Nature* **203**, No. 4943, 393.

Piggot, C. S. and Urey, W. D. (1942). *Amer. J. Sci.* **240**, 93.

Pinson, W. H., Schnelgler, C. C. and Beiser, E. (1962). 10th Ann. Rep. Progr. U.S. Atomic Energy Commission Contract AT(30–1) 1381, 19.

Plesset, M. and Lather, A. (1960). *Proc. nat. Acad. Sci., Wash.* **46**, 1232.

Poleyava, N. I., Murina, G. A. and Kazakov, G. A. (1961). *Ann. N.Y. Acad. Sci.* **91**, Art. 2, 298.

Poplavko, E. M., Marchakova, I. D. and Zak, S. Sh. (1962). *C.R. Acad. Sci. U.R.S.S.* **146**, 433.

Posanker, A. M. and Foreman, B. M. (1961). *J. inorg. nucl. Chem.* **16**, 323.

Powell, J. L. and Hurley, P. M. (1963). *Geol. Soc. Amer. Ann. Meeting Abstr.*

Powell, J. L., Hurley, P. M. and Fairbairn, H. W. (1962). *Nature* **196**, No. 4859, 1085.

Powell, J. L., Faure, G. and Hurley, P. M. (1964). *Trans. Amer. geophys. Un.* **45**, No. 1, 114.

Price, P. B. and Walker, R. M. (1962a). *J. appl. Phys.* **33**, No. 12, 3400.

Price, P. B. and Walker, R. M. (1962b). *J. appl. Phys.* **33**, No. 12, 3407.

Price, P. B. and Walker, R. M. (1963). *J. geophys. Res.* **68**, No. 16, 4847.

Quinn, A. W., Jaffe, H. W., Smith, W. L. and Waring, C. L. (1957). *Amer. J. Sci.* **255**, 547.

Rankama, K. (1954). "Isotope Geology". Interscience, New York.

Rankama, K. (1963). "Progress in Isotope Geology". Wiley, New York.

Rausch, W. and Schmidt, W. (1960). Fachausschuss Kernphysik Tagung in Heidelberg.

Reade, T. M. (1893). *Geol. Mag.* **10**, 97.

Reed, G. W., Kigoshi, K. and Turkevich, A. (1960). *Geochim. Cosmochim. Acta* **20**, 251.

Reuterswärd, C. (1951). *Ark. Fys.* **3**, 53.

Reuterswärd, C. (1956). *Ark. Fys.* **11**, 1.

Revelle, R. and Suess, H. E. (1957). *Tellus*, No. 1, **9**, 18.

Reynolds, J. H. (1956). *Rev. sci. Instrum.* **27**, 928.

Reynolds, J. H. (1957). *Geochim. cosmochim. Acta* **12**, 177.

Reynolds, J. H. (1960). *Phys. Rev. Letters* **4**, 8, 351.

Reynolds, J. H. (1963). *J. Geophys. Res.* **68**, 2939.

Reynolds, J. H., Merrihue, C. M. and Pepin, R. O. (1962). *Bull. Amer. phys. Soc.* **7**, 35.

Richards, J. R. (1962). *Microchim. Acta* **4**, 620.

Richards, J. R. and Pidgeon, R. T. (1963). *Geol. Soc. Aust.* **10**, Pt. 2, 243.

Richards, T. W. and Lembert, M. E. (1914). *J. Amer. chem. Soc.* **36**, 1329.

Robinson, J. W. (1962). *Ind. Chem.* May-July, 52.

Robinson, S. C., Loveridge, W. D., Rimsaite, J. and van Peteghem, J. (1963). *Canad. Min.* **17**, Pt. 3, 533.

Rodden, C. J. (1950). "Analytical Chemistry of the Manhattan Project". Nat. Nucl. Energy Ser. McGraw-Hill, New York.

Rona, E., Gilpatrick, L. O. and Jeffrey, L. M. (1956). *Trans. Amer. geophys. Un.* **37**, 697.

Rosholt, J. N. (1957). *Analyt. Chem.* **27**, 1398.

Rosholt, J. N. and Dooley, J. R. (1960). *Analyt. Chem.* **32**, 1093.

Rosholt, J. N., Emiliani, C., Geiss, J., Koczy, F. F. and Wangersky, P. J. (1961). *J. Geol.* **69**, No. 2, 162.

Rubin, M. (1961). *Radiocarbon* **3**, 86.

Russell, R. D. (1963). *In* "Earth Science and Meteorites", ed. by J. Geiss and E. D. Goldberg, p. 44. North Holland Publ. Co., Amsterdam.

Russell, R. D. and Ahrens, L. H. (1957). *Nature* **179**, 92.

Russell, R. D. and Allan, D. W. (1956), *Royal Astron. Soc. Mth. Notices*, Geophys. Suppl. **7**, 80.

Russell, R. D. and Farquhar, R. M. (1957). *Trans. Amer. geophys. Un.* **38**, 557.

Russell, R. D. and Farquhar, R. M. (1960a). *Geochim. cosmochim. Acta* **19**, 41.

Russell, R. D. and Farquhar, R. M. (1960b). "Lead Isotopes in Geology", p. 243. Interscience, New York.

Russell, R. D., Farquhar, R. M. and Hawley, J. E. (1957). *Trans. Amer. geophys. Un.* **38**, 557.

Russell, R. D., Kollar, F. and Ulrych, R. J. (1961). *J. Geophys. Res.* **66**, 1495.

Russell, R. D., Farquhar, R. M., Cumming, G. L. and Wilson, W. T. (1954). *Trans. Amer. geophys. Un.* **35**, 301.

Rutherford, E. and Soddy, F. (1902). *Trans. chem. Soc.* **81**, 937; *Phil. Mag.* (6) **4**, 370, 569.

Salmon, L. (1957). A.E.R.E., C/M.323. Rept.

Samuelson, O. (1953). "Ion Exchanges in Analytical Chemistry". Wiley, New York.

Sano, S. and Nakai, J. (1961). *J. Atomic Energy Soc. Japan* **3**, No. 4, 288.

Sardarov, S. S. (1961). *Geokhimiya* No. 1, 30.

Schaeffer, O. A. (1960). *Physics Today* **18**, 22,

Schaeffer, O. A. (1962). *Ann. Rev. phys. Chem.* **13**, 151.

Schaeffer, O. A., Stoenner, R. W. and Bassett, W. A. (1961). *Ann. N.Y. Acad. Sci.* **91**, Art. 2, 317.

Schreiner, G. D. O. (1958). *Proc. roy. Soc.* **A245**, 112.

Schuler, H. and Jones, E. C. (1932). *Naturwissenschaften* **20**, 171.

Schumacher, E. (1956). *Z. Naturf.* **11a**, 206.

Sharkey, A. G., Robinson, C. F. and Friedel, R. A. (1959). *In* "Advances in Mass Spectrometry", p. 193. Pergamon Press, London.

Shaw, D. M. (1957). *Econ. Geol.* **52**, 570.

Shields, W. R., Garner, E. L., Hedge, C. E. and Goldich, S. S. (1963). *J. geophys. Res.* **68**, No. 8, 2331.

Shillibeer, H. A. and Russell, R. D. (1954). *Canad. J. Phys.* **32**, 681.

Shillibeer, H. A., Russell, R. D., Farquhar, R. M. and Jones, E. A. W. (1954). *Phys. Rev.* **94**, 1793.

Shima, M., Gross, W. H. and Thode, H. G. (1963). *J. geophys. Res.* **68**, No. 9, 2835.

Signer, P. and Nier, A. O. (1960). *J. geophys. Res.* **65**, 2947.

Silk, E. C. H. and Barnes, R. S. (1959). *Phil. Mag.* **4**, 970.

Sill, C. W. and Willis, C. P. (1962). *Anal. Chem.* **34**, 954.

Silver, L. T. (1960). *Bull. geol. Soc. Amer.* **71**, 1973.

Silver, L. T. (1963). *In* "Radioactive Dating", p. 279. IAEA, Vienna.

Silver, L. T. and Deutsch, S. (1963). *J. Geol.* **71**, No. 6, 721.

Simpson, J. A. (1948). *Phys. Rev.* **73**, 1389.

Slawson, W. F. and Austin, C. F. (1960). *Nature* **187**, 400.

Slawson, W. F. and Austin, C. F. (1962). *Econ. Geol.* **57**, 21.

Slawson, W. F. and Nackowski, M. P. (1959). *Econ. Geol.* **54**, 1543.

Smales, A. A. (1949). *Ann. Rep. chem. Soc.* **46**, 285.

Smales, A. A. (1952). *Analyst* **77**, No. 920, 778.

Smales, A. A. (1955). *Geochim. cosmochim. Acta* **8**, 300.

Smales, A. A. and Webster, R. K. (1957). *Geochim. cosmochim. Acta* **11**, 139.

Smales, A. A., Hughes, T. C., Mapper, D., McInnes, C. A. J. and Webster, R. K. (1964). *Geochim. cosmochim. Acta* **28**, 209.

Smit, F. and Gentner, W. (1950). *Geochim. cosmochim. Acta* **1**, 22.

Smith, J. L. (1871). *Amer. J. Sci.* **1**, 269.

Smitheringale, W. G. and Jensen, M. L. (1962). *Geochim. cosmochim. Acta* **27**, 1183.

Smulikowski, I. (1954). *Polska Akad. Nauk. Kom. Geol. Arch. Miner.* **18**, 21.

Smythe, W. R. and Hemmendinger, A. (1937). *Phys. Rev.* **51**, 178.

Soddy, F. and Hyman, H. (1914). *Trans. chem. Soc.* **105**, 1402.

Sorensen, A. H. (1963). *Econ. Geol.* **58**, No. 7, 1071.

Stanton, R. L. (1955a). *Econ. Geol.* **50**, 681.

Stanton, R. L. (1955b). *Aust. J. Sci.* **17**, 173.

Stanton, R. L. (1960). *N.Z. J. Geol. Geophys.* **3**, No. 3, 375.

Stanton, R. L. and Russell, R. D. (1959). *Econ. Geol.* **54**, 588.

Starik, I. E. (1961). "Yadernaya Geokhronologiya". Akademiya Nauk SSSR, Moscow.

Starik, I. E., Starik, F. E. and Petrayaev, E. P. (1955). *Byull. Komissii Opredelen Absolyut Vozrasta Geol. Formatsii, Akad. Nauk SSSR* **1**, 29.

Starik, I. E., Sobotovich, E. V., Lovtsyus, G. P., Lovtsyus, A. V. and Avdzeiko, G. V. (1957). *Geokhimiya* **7**, 584.

Starik, I. E., Sobotovich, E. V., Lovtsyus, G. P., Lovtsyus, A. V., and Shats, M. M. (1959). *Radiokhimiya* **1**.

Starik, I. E., Sobotovich, E. V., Lovtsyus, G. P., Shats, M. M. and Lovtsyus, A. V. (1960). *Dokl. Akad. Nauk SSSR* **134**, 555.

Stern, T. W. and Rose, H. J. Jr. (1961). *Amer. Min.* **46**, 606.

Stevens, C. M. (1953). *Rev. sci. Instrum.* **24**, 148.

Stieff, L. R. and Stern, T. W. (1956). U.S. Geol. Survey Prof. Paper, 300, p. 549.

Stieff, L. R. and Stern, T. W. (1961). *Geochim. cosmochim. Acta* **22**, Nos. 2–4,176.

Stieff, L. R., Stern, T. W., Oshiro, S. and Senftle, F. E. (1959). U.S. Geol. Surv. Prof. Paper 334A, p. 1.

Stoenner, R. W. and Zähringer, J. (1958a). *Geochim. cosmochim. Acta* **1**, 22.

Stoenner, R. W. and Zähringer, J. (1958b). *Geochim. cosmochim. Acta* **14**, 40.

Strassmann, F. and Walling, E. (1938). *Ber. dtsch. chem. Ges.* **B71**, 1.

Straus, H. A. (1941). *Phys. Rev.* **59**, 430.

Strutt, R. J. (1905). *Proc. roy. Soc.* **A76**, 88.

Strutt, R. J. (1908). I. *Proc. roy. Soc.* **A81**, 272.

Strutt, R. J. (1909). II. *Proc. roy. Soc.* **A83**, 96.

Strutt, R. J. (1910). III. *Proc. roy. Soc.* **A83**, 298.

Suess, H. E. (1948). *Z. Phys.* **125**, 386.

Suess, H. E. (1954). *Science* **120**, 5.

Suess, H. E. (1955). *Science* **122**, 415.

Takahaski, J. and Yagi, T. (1929). *Econ. geol.* **24**, 838.

Tatsumoto, M. and Patterson, C. (1963). "Earth Sciences and Meteoritics", p. 74. North Holland Publ. Co., Amsterdam.

Taylor, H. P. and Epstein, S. (1963). *J. Petr.* **4**, Pt. 1, 51.

Thode, H. G., Kleerekoper, N. and McElcheran, A. (1951). *Research* **4**, 581.

Thode, H. G., Macnamara, J. and Collins, C. B. (1949). *Canad. J. Res.* **27B**, 361.

Thode, H. G., Macnamara, J. and Fleming, W. H. (1953). *Geochim. cosmochim. Acta* **3**, 235.

Thode, H. G., Monster, J. and Dunford, H. B. (1961). *Geochim. cosmochim. Acta* **25**, 150.

Thode, H. G., Wanless, R. and Wallauch, R. (1954). *Geochim. cosmochim. Acta* **5**, 228.

Thomson, P. F., Perry, M. A. and Byerly, W. M. (1949). *Analyt. Chem.* **21**, 1236.

Thomson, J. J. (1905). *Phil. Mag.* (6) **10**, 584.

Tilley, D. R. and Madansky, L. (1959). *Phys. Rev.* **116**, 413.

Thurber, D. L. (1962). *J. geophys. Res.* **67**, No. 11, 4518.

Tilton, G. R. (1951). U.S. AEC. Rept. 3182.

Tilton, G. R. (1956). *Trans. Amer. Geophys. Un.* **32**, No. 2, 224.

Tilton, G. R. (1960). *J. geophys. Res.* **65**, No. 9, 2933.

Tilton, G. R. and Davis, G. L. (1959). *In* "Researches in Geochemistry", ed. by P. H. Abelson, p. 190, Wiley, New York.

Tilton, G. R. and Nicolaysen, L. O. (1955). *Geochim. cosmochim. Acta* **11**, 28.

Tilton, G. R. and Patterson, C. (1956). *Trans. Amer. geophys. Un.* **37**, 361.

Tilton, G. R. and Reed, G. W. (1960). Summer Course on Nuclear Geology, Comitato Nazionale per L'Energia Nucleare, p. 367.

Tilton, G. R., Aldrich, L. T. and Inghram, M. G. (1954). *Analyt. Chem.* **26**, 894.

Tilton, G. R. and Reed, G. W. (1963). *In* "Earth Science and Meteorites", ed. by J. Geiss and E. D. Goldberg, p. 34. North Holland Publ. Co., Amsterdam.

Tilton, G. R., Patterson, C., Brown, H., Inghram, M. G., Hayden, R. J., Hess, D. and Larsen, E. S. (1955). *Bull. geol. Soc. Amer.* **66**, 1131.

Tilton, G. R., Davis, G. L., Wetherill, G. W. and Aldrich, L. T. (1957). *Trans. Amer. geophys. Un.* **38**, 360.

Tomlinson, R. H. and Das Gupta, A. K. (1953). *Canad. J. Chem.* **31**, 909.

Trask, P. D. (1955). Soc. Econ. Pal. Mineral Special Pub. No. 1. Tulsa, Oklahoma.

Turekian, K. K. and Kulp, J. L. (1956). *Geochim. cosmochim. Acta* **10**, 245.

Turnbull, A. H., Barton, R. S. and Riviere, J. C. (1962). "An Introduction to Vacuum Technique". Newnes, London.

Ulrych, T. J. (1964). *Geochim. cosmochim. Acta* **28**, 1389.

Ulrych, T. J. and Russell, R. D. (1964). *Geochim. cosmochim. Acta* **28**, 344.

Urey, H. C. (1951). "The Planets". Yale University Press.

Urey, H. C. and Craig, H. (1953). *Geochim. cosmochim. Acta* **4**, 36.

Vincent, E. A. (1960). *In* "Methods in Geochemistry", ed. by A. A. Smales and L. R. Wager, p. 33. Interscience, New York.

Vinogradov, A. P. (1956). *Geokhimiya* **5**, 427.

Vinogradov, A. P. (1958). "Radioisotopes in Scientific Research", Vol. 11, p. 581. Pergamon Press, London.

Vinogradov, A. P. and Tugarinov, A. I. (1961). *Ann. N.Y. Acad. Sci.* **91**, Art. 2, 500.

Vinogradov, A. P., Chupakhim, M. S. and Grinenko, V. A. (1957). *Geokhimiya* **3**, 183.

Vinogradov, A. P., Tarasov, L. S., and Zykov, S. I. (1957). *Geokhimiya* **1**, 1.

Vinogradov, A. P., Tarasov, L. S. and Zykov, S. I. (1959). *Geokhimiya* **7**, 689.

Vinogradov, A. P., Zykov, S. I. and Tarasov, L. S. (1958). *Geokhimiya* No. 6, 653.

Volchok, H. L. and Kulp, J. L. (1957). *Geochim. cosmochim. Acta* **11**, 219.

Voshage, H. and Hintenberger, H. (1960). *Nature* **185**, 88.

de Vries, H. L. (1956). *Nuclear Physics* **1**, 477.

de Vries, H. L. (1957). *Appl. Sci. Res.* Section B6.

de Vries, H. L. (1959). *In* "Researches in Geochemistry", ed. by P. H. Abelson, p. 164. Wiley, New York.

de Vries, H. L. and Barendsen, G. W. (1952). *Physica* **18**, 652.

de Vries, H. and Oakley, K. P. (1960). *Nature* **184**, 224.

Wager, L. R. and Brown, G. M. (1960). *In* "Methods in Geochemistry", ed. by A. A. Smales and L. R. Wager, p. 4. Interscience, New York.

Wager, L. R. and Deer, W. A. (1939). *Meded. Grønland* **105**, No. 4, 353.

Wahl, W. (1940). *In* "Rept. of the Committee on the Measurement of Geologic Time", 1940–1941, p. 65. Nat. Res. Coun. U.S.A.

Wahl, W. (1942). *Geol. Rdsch.* **32**, 550.

Wampler, J. M. and Kulp, J. L. (1962). A. F. Buddington Volume *Bull. geol. Soc. Amer.* **105**.

Wampler, J. M. and Kulp, J. L. (1964). *Geochim. cosmochim. Acta* **28**, 1419.

Wänke, H. (1960). "Summer Course on Nuclear Geology". Varenna.

Wänke, H. and König, H. (1959). *Z. Naturf.* **14a**, 860.

Waring, C. L. and Mela, H. (1953). *Analyt. Chem.* **25**, 423.

Waring, C. L. and Worthing, H. (1953). *Amer. Min.* **38**, 827.

Warshaw, C. M. (1957). "The Mineralogy of Glauconite". Unpublished Ph.D. thesis, Pennsylvania State University.

Washington, H. S. (1930). "The Chemical Analysis of Rocks". Wiley, New York.

Wasserburg, G. J. (1961). *Ann. N.Y. Acad. Sci.* **91**, Art. 2, 583.

Wasserburg, G. J. (1963). *J. geophys. Res.* **68**, No. 16, 4823.

Wasserburg, G. J. and Birch, R. (1958). *Geochim. cosmochim. Acta* **15**, No. 112, 157.

Wasserburg, G. J. and Hayden, R. J. (1954). *Phys. Rev.* **93**, 645.

Wasserburg, G. J. and Hayden, R. J. (1955). *Geochim. cosmochim. Acta* **7**, 51.

Wasserburg, G. J., Hayden, R. J. and Jensen, K. J. (1956). *Geochim. cosmochim. Acta* **10**, 153.

Webster, R. K. (1958). *In* "Advances in Mass Spectrometry", p. 103. Pergamon Press, London.

Wedepohl, K. H. (1956). *Geochim. Cosmochim. Acta* **10**, 69.

von Weiszäcker, C. F. (1937). *Phys. Z.* **38**, S, 623.

Wetherill, G. W. (1956). *Trans. Amer. geophys. Un.* **37**, 320.

Wetherill, G. W. (1957). *Science* **126**, 545.

Wetherill, G. W. (1963). *J. geophys. Res.* **68**, No. 10, 2957.

Wetherill, G. W., Tilton, G. R., Davis, G. L. and Aldrich, L. (1956). *Geochim. cosmochim. Acta* **9**, No. 516, 292.

Wetherill, G. W., Wasserburg, G. J., Aldrich, L. T. and Tilton, G. R. (1956). *Phys. Rev.* **103**, No. 4, 987.

Whitefield, J. M., Rogers, J. J. W. and Adams, J. A. S. (1959). *Geochim. Cosmochim. Acta* **17**, 248.

Whiting, F. B. (1950). Ph.D. Thesis, Massachusetts Institute of Technology.

Wickman, F. E., Blomquist, N. G., Geiger, P., Parwel, A., v. Ubisch, H. and Welin, E. (1963). *Ark. Miner. Geol.* **3**, No. 11, 193.

Willard, H. H., Merritt, L. L. and Dean, J. A. (1958). "Instrumental Method of Analysis", 3rd Ed. Van Nostrand, Princeton.

Williams, A. F. (1952). *Analyst* **77**, 297.

Williams, H. S. (1893). *J. Geol.* **1**, 283.

Wilson, C. I. and Wilson, D. W. (1959). "Comprehensive Analytical Chemistry". Elsevier, Amsterdam.

Wilson, J. T. (1949). *Canad. Min. Metal. Bull.* 231.

Wilson, J. T. (1963). *Canad. J. Phys.* **41**, 863.

Winchester, J. W. (1961). *Analyt. Chem.* **33**, 1007.

Winsberg, L. (1956). *Geochim. cosmochim. Acta* **9**, 183.

Witney, P. R. (1962). *AEC Rep.* AT(30-1).

Witney, P. R. and Hurley, P. M. (1964). *Geochim. cosmochim. Acta* **28**, No. 4, 425.

Woldring, M. G. (1953). *Anal. Chem. Acta* **8**, 150.

Wood, L. and Libby, W. F. (1964). "Isotopic and Cosmic Chemistry", ed. by H. Craig, S. L. Miller and G. J. Wasserburg, p. 205. North Holland Publ. Co., Amsterdam.

Yagoda, H. (1949). "Radioactive Measurements with Nuclear Emulsions". Wiley, New York.

Yoe, J. H., Will, F. and Black, R. A. (1953). *Analyt. Chem.* **25**, 1200.

Zähringer, J. (1960). *Geol. Rsch.* **49/1**, 224.

Zähringer, J. (1964). *Ann. Rev. Astron. Astrophys.* **2**, 121.

Zartman, R. E. (1963). U.S. Nat. Acad. Sci. Nat. Res. Coun. Publ. 1075. Nucl. Sci. Ser. Rep. No. 38, 43.

Zeuner, F. E. (1952). "Dating the Past". Methuen, London.

Zeuner, F. E. (1958). "An Introduction to Geochronology", 4th Ed. Methuen, London.

Zhirov, K. K. and Zykov, S. I. (1958). *Er. Gegr. Geol.* No. 1, 258.

Zhirov, K. K., Zykov, S. I. and Stupnikova, N. I. (1957). *Geokhimiya* **5**, 430.

AUTHOR INDEX

A

Abbey, S., 62, *229*

Abelson, P. H., 83, *229*

Adams, J. A. S., ix, xii, xiii, 69, 133, 185, *229, 235, 240, 245*

Agerter, S., 45, *234*

Ahrens, L. H., ix, xi, 48, 70, 81, 82, 83, 137, 139, 142, *229, 237, 242*

Aitkin, K. L., 21, *241*

Akishin, P. A., 172, *229*

Aldrich, L. T., 15, 21, 47, 48, 50, 51, 52, 83, 86, 87, 140, 145, 146, 195, *229, 232, 235, 244, 245*

Alkemade, C. Th., 59, *235*

Allan, D. W., 180, 188, *229, 242*

Allen, J. S., 25, *229*

Allsop, H. L., 96, 99, 100, 102, *229*

Alpert, D., 53, *229*

Alpher, R. A., 180, *229*

Amirkhanoff, K., 58, 76, 77, *229*

Anders, E., 108, 109, 111, 214, 220, *229*

Anderson, E. C., 32, 34, 36, 40, 41, 42, *229, 235*

Antweiler, J. C., 202, *231*

Arnold, J. R., 36, 40, 41, 43, 213, *229, 235*

Arrhenius, G., 208, 212, *229*

Aston, F. W., 7, 17, *229*

Ault, W. U., 204, 205, *229, 238*

Austin, C. F., 199, *243*

Avdzeiko, G. V., 187, *243*

B

Baadsgaard, H., 73, 76, 77, 145, *229, 233, 235*

Bachus, M. M., 209, *230*

Badlor, S. T., 121, *230*

Baethe, N., 105, *237*

Bainbridge, K. T., 49, 81, *230*

Ballaria, C., 43, *230*

Baranov, V. I., 210, *230*

Baranovskaya, N. V., 48, 74, *234, 238*

Barendsen, G. W., 37, 41, *230, 245*

Barnard, G. P., 17, *230*

Barnes, R. S., 153, *243*

Barrell, J., 7, 8, *230*

Bartnitsky, E. N., 58, 76, 77, *229*

Barton, R. S., 27, *244*

Basitova, S. M., 122, *230*

Bassett, W. A., 72, *240, 242*

Bate, G. L., 142, 173, 190, *230, 238*

Bauer, C. A., 45, *230*

Bazan, F., 213, *234*

Beard, G. B., 87, *230*

Begemann, F. V., 149, 213, *230*

Beiser, E., 215, *243*

Belcher, R., 59, *230*

Bell, J. D., 117, *240*

Benson, S. W., 173, *230*

Bentley, P. G., 22, *230*

Beukelman, T. E., 62, *230*

Beynon, J. H., 17, *230*

Bien, G. S., 45, *230*

Bikerman, M., 223, *231*

Birch, F., 47, *230*

Birch, R., 47, *246*

Bird, J., 35, *230*

Bishop, J., 22, 137, *230*

Black, R. A., 129, *246*

Blomquist, N. G., 200, *246*

Bloxham, T. W., 133, *230*

Bolton, E. T., 83, *229*

Boltwood, B. B., 5, *230*

Bowie, S. H. U., 136, *230*

Brambelle, M. N., 208, *229*

Brandt, S. B., 58, 76, 77, *229*

de Breuck, W., 150, *231, 243*

Brewer, A. K., 22, 48, 85, *230*

Bright, H. A., 129, *235*

Broecker, W. S., 35, 142, 149, 150, *230, 238*

Brongniart, P., 1, *231*

Brown, E. W., 6, 137, *237*

Brown, G. M., 90, *245*

Brown, H., 145, 171, 187, 195, 198, 220, *241, 244*

Brown, R. M., 39, *234*

Buck, K. L., 202, *231*

Bullard, E. C., 188, *230*

SUBJECT INDEX

A

α-Particle emitters, 4, 7, 126, 127, 128, 130, 132, 134, 160, 165
Achondrites
  K–Ar age of, 217
  Rb–Sr age of, 108
Acid-washing experiments, 143
Alpha-helium dating, 5, 6, 7, 163
Analysis by isotope dilution, 10
  argon, 57
  general equation, 13
  isobaric interference, 14
  optimum conditions, 14
  scope of method, 14
  sensitivity precision, 15
  tracer calibration, 11
Anomalous leads, 189
Argon
  diffusion in minerals, 76, 77
  excess in minerals, 71
  extraction from minerals, 73, 79
  in biotite, 72, 77
  in feldspars, 74, 77
  in fluorite, 75
  in glauconite, 74
  in illite, 75
  in limestones, 75
  in meteorites, xi, 72
  in pyroxenes, 71
  in sanidine, 73, 77
  in slates, 73
  in whole rocks, 72, 75
  loss, 76, 79
  measurement of, 52, 56
  in silts, 74
  in volcanics, 72
  in volcanic ash, 73
  retentivity in minerals, 73, 79
Argon dating, 47
  basic rocks, 72
  bentonite clay, 73, 77
  decay constants, 49, 50, 51
  decay scheme, 48
  excess argon in pyroxenes, 72
  feldspars, 77

fluorite, 75
fossil bones, 75
glauconite, 74
igneous rocks, 71
sediments, 73
separation from minerals, 55
Argon isotopes
  ^{36}Ar, as tracer, 58, 71
  ^{37}Ar, as tracer, 71
  ^{38}Ar, as tracer, 58
  ^{39}Ar, as tracer, 58
  ^{40}Ar, determination of content, 58
    in minerals, 48
  ^{41}Ar, as tracer, 70
  in meteorites, 217
  isotope dilution, 57, 58
  metamorphic events, 71, 79
  natural abundances, 47, 49
  potassium argon dating, 47
  production in igneous rocks, 47
  radioactive, 47
Autoradiography, 134
  α-distribution in G1, 135
  α-distribution in rocks, minerals, 135, 136
  α-sensitive emulsions, 134
  β-sensitive emulsions, 136

B

β-Emitters, 4, 126, 127, 128
B-type lead, 189, *see* Anomalous leads
Basalt
  Pb content of olivine nodules, viii
  Ca content, viii
  K/Ar ratio in, 72, 73, 223
  Pb isotopes in, 183, 185
  Pb/U ratio in, 179
  Rb content of, viii
  S isotopes in, 206
  Sr content of, viii
  Sr isotopes in, 109, 111, 118
  Th content, viii
  U content, viii
Base Roi Baudouin, S. Pole, 151